Principles and Pra[ctice of]
Dressing and Fu[nishing]

William E. Austin

Alpha Editions

This edition published in 2024

ISBN 9789362516176

Design and Setting By
Alpha Editions
www.alphaedis.com
Email - info@alphaedis.com

Contents

PREFACE

The great increase in the use of furs during the past few decades has caused the fur dressing and dyeing industry to rise from relative insignificance to considerable importance as a branch of applied chemistry. The past eight years, moreover, have witnessed the virtual transference of the leadership in the dressing and dyeing of furs from Europe to America, and in the quality and variety of products, the domestic industry is now in every way the equal of, and in many respects superior to the foreign. The great bulk of American furs which formerly were sent to Leipzig, Paris or London to be dressed and dyed, are now being dressed and dyed in this country.

In spite of these facts, very little is generally known about the nature and manner of the work constituting the dressing and dyeing of furs. Even among members of other branches of the fur trade, there is very little accurate information on the subject. Real knowledge concerning fur dressing and dyeing is possessed only by those actually engaged in the industry. The interest and efforts of scientists and technologists have been enlisted to only a small extent in the technical development of the industry. The reason for this may be attributed to two related causes: first, the almost monastic seclusion in which fur dressers and dyers, particularly the latter, conducted their operations, and even to-day the heavy cloud of mystery is being dispelled but very slowly; and second, as a consequence of the first, the lack of any reliable literature on the subject. Of the few books which have been written on the industry of fur dressing and fur dyeing (all of them either German or French), most are hopelessly out of date, or contain no trustworthy data; or, if they do have real merit, they cannot be obtained. Numerous articles in the technical journals are of interest, but they contain very little information of value.

This work is intended for a two-fold purpose: first, that it may serve as a text-book for those who expect to make fur dressing and dyeing their vocation. The fundamental principles upon which the industry is based are discussed in the light of the most recent chemical and technical developments, and the most important operations are treated fully and systematically, and are illustrated with practical examples.

Secondly, as a practical handbook for the worker in the fur dressing and dyeing plant. The latest factory processes and methods are described, and numerous working formulas given. The formulas are all such as have been successfully used on a large scale, and give satisfactory results when applied under the proper conditions.

In addition, it is believed that the book will prove of interest to chemists and other students of industrial chemistry, since it will be an introduction into a field of applied chemistry, about which very little is known to those outside of the industry.

Thanks are due to Dr. L. A. Hausman, of Cornell University, for material used in Chapter II; to Dr. E. Lesser of the American Dyewood Company, for information and assistance on the subject of Vegetable Dyes; to the Gaskill Chemical Corp., American Aniline Products, Inc., the Cassella Company, and the Franklin Import & Export Co., for information about their products in connection with the chapter on Oxidation Colors; to F. Blattner, Fletcher Works, Inc., S. M. Jacoby Co., Proctor & Schwartz, Inc., Reliable Machine Works, Seneca Machine & Tool Co., Inc., and the Turner Tanning Machinery Co., for the use of the cuts of the various machines.

WILLIAM E. AUSTIN.

NEW YORK, May, 1922.

CHAPTER I
FURS AND THEIR CHARACTERISTICS

Furs have in general two uses: as the goods which constitute the basis of the furrier's art, and as the source of material for the hat manufacturer. In the latter case, only the hair part of the fur is utilized in the hat trade for the production of felt, the skin being either made into leather, or used as the raw material for making high-grade glue and gelatine. It is the furrier, therefore, who uses the great bulk of furs, and requires them to be dressed and dyed.

In discussing the dressing and the dyeing of furs, there are, broadly speaking, two fundamental subjects to be considered: first, the raw materials employed, which are, of course, the skins or pelts as they come from the trapper. (Other substances used in fur dressing and dyeing are accessories, and will be studied in connection with the processes.) Second, all those operations, physical and chemical, manual and mechanical, to which the raw skins have to be subjected in order to obtain the finished fur, ready for use by the furrier.

Next to the inherent qualities of the fur skin, the future value of a fur in a manufactured garment depends largely on the dressing and dyeing it receives. It is in these operations that the beauty of the fur can be brought out to its fullest degree, and if possible, enhanced, or the attractive features can be marred or destroyed, and the fur rendered quite worthless. Therefore, it is quite essential for the fur dresser and the fur dyer in addition to the technical knowledge and experience which are the fundamental requisites of the industry, also to have more than a superficial familiarity with the various kinds of furs. In fact, an accurate knowledge of the nature and chief characteristics of furs in general, and of the individual classes, in particular, is almost indispensable to obtain the best results. The habits and habitats of the various fur-bearing animals are factors which largely determine the constitution of the fur, and the nature of the skin. There are as many different kinds of fur hair, with as many different kinds of skin bearing the hair, as there are classes of furs. The methods of dressing, and often, if the furs are to be dyed, the manner of dyeing, are determined by the nature of these component parts of furs. Various chemicals affect furs in widely different ways. The divergence with regard to the physical and chemical properties of the classes of furs is such as to make almost imperative a detailed knowledge of the typical members of the many groups of commercial furs.

To be sure, there are many engaged in the dressing and dyeing of furs, who never made a formal study of this phase of the industry, but acquired their knowledge empirically, and are apparently quite successful. It must not be denied, that practise and experience, as in every field of enterprise, are essential to obtaining the best results. But the time and cost of acquiring this precious experience can be considerably reduced by systematically studying the important characteristics and properties of furs. These will be treated briefly, but in sufficient detail to form a basis for discussing the operations of dressing and dyeing.

Fur-bearing animals are mammals whose skins are used in the manufacture of fur garments and other fur wearing apparel. The skin, when it is removed from the animal is called a pelt, or sometimes, in the case of large animals, a hide. The pelt, after having been dressed and dyed, is called a fur, the skin part being referred to as the leather, and the hair as the pelage. However, this terminology is not strictly adhered to in practise, and the various terms are often employed interchangeably.

The various fur-bearing animals differ considerably in the characteristics of the furs they yield. With few exceptions, notably beaver and Alaska red fox, the depth of shade increases as the habitat of the animal species is nearer the equatorial regions. There seems to be a direct relationship between the intensity of color of the pelt, and the distance from, or proximity to the polar, or the torrid regions. Thus, white mammals, such as polar bear, ermine, white or Siberian hare, are found only in the northern lands. An exception is the sheep, which, due to its domestic nature, can be found in almost all parts of the civilized world. Tropical animals on transportation to colder climates, have been known to become lighter-haired when adapted to their new environment. The skins of animals living in dense woods or forests, are generally of a deeper color than in animals living in more open territory. As a general rule, fur-bearing animals have darker hair on the back than on the sides and belly. The badger, hamster, ratel and panda are exceptions having the darker hair on the belly and sides, and the lighter hair on the back. With regard to the intensity of color, the skunk has the blackest fur, although some domestic cats are also quite black. Other animals whose fur is nearly black, are the black bear, and the black fox, which is a variety of the silver fox, but the color is often of a brownish shade. The colors which predominate among animals of the fur-bearing variety, are white, black, brown, and grey. Less common are yellow shades, and those known as blue.

The quality of the fur on all mammals improves with cold, and animals living at greater altitudes, with correspondingly lower temperatures, have thicker and finer hair than those living nearer sea-level. A cold winter generally produces fur of high quality and fine color, a mild winter may cause the hair to be inferior. In all climates, animals found in dense woods, have fur which is deeper, silkier, thicker, and glossier than that of animals living in the open. Animals inhabiting inland lakes and rivers, have finer and softer hair than those living near the coast or land exposed to sea winds. In general, the hair of animals of the cold regions is short, fine, soft, and downy, while the hair of animals of warmer lands, is longer, stiffer, and harder.

Both the quality and color of the fur vary with the age of the animal. The young usually have a thicker coat of fur than adults, but the hair is too soft, and the skin generally too tender to be fit for use. In certain cases, particularly the baby lambs, very young skins are especially prized, and eagerly sought, but extraordinary care has to be exercised in working with them. Fur is at its best when the animal is between one and two years old. After this age, the fur becomes coarse and scraggy. The animal attains its fullest growth of hair usually in the height of winter, and the fur is best between then and very early spring. Before mid-winter the hair is short and thin, and in the spring it begins to shed, and will continue to fall out even in the dressed fur. The color of the hair also becomes lighter with age, and the new growth which generally comes in the fall is darker than the old coat.

Different members of the same species, will, other factors such as age and season being equal, vary as to color and quality. There may even be several different color phases of the same species of animal, such as the cross fox and the silver fox, both of which are of the same genus as the red fox; black muskrats are of the same class as the brown variety, etc. The individual pelt likewise presents many variations in color and nature of the hair. In some parts, the hair is thicker and softer than others, and the color varies in intensity and shade throughout the different sections of the skin.

Furs do not have differences confined to the hair part only; the leather also presents considerable variation among the different fur-bearing animals, especially in regard to the weight and thickness. The durability of furs, relatively considered under similar conditions of wear, also varies widely. In the following table the relative durability of dressed furs, and in certain instances also dyed furs, otter being taken as standard, is given, as well as the weight in ounces per square foot of skin of these furs.

Name of Fur	Durability Otter = 100	Wt. in oz. per sq. ft.
Astrachan	10	3
Bear, brown or black	94	7

Beaver, natural	90	4
Beaver, plucked	85	$3\frac{7}{8}$
Chinchilla	15	$1\frac{1}{2}$
Civet cat	40	$2\frac{3}{4}$
Coney	20	3
Ermine	25	$1\frac{1}{4}$
Fox, natural	40	3
Fox, dyed black	25	3
Genet	35	$2\frac{3}{4}$
Goat	15	$4\frac{1}{8}$
Hare	05	$2\frac{1}{4}$
Krimmer	60	3
Kolinsky	25	3
Leopard	75	4
Lynx	25	$2\frac{3}{4}$
Marten, Baum natural	65	$2\frac{3}{4}$
Marten, Baum blended	45	$2\frac{3}{4}$
Marten, Stone natural	45	$2\frac{7}{8}$
Marten, Stone dyed	35	$2\frac{7}{8}$
Mink, natural	70	$3\frac{1}{4}$
Mink, dyed	35	$3\frac{1}{4}$
Mink, Jap	20	3
Mole	07	$1\frac{3}{4}$
Muskrat	45	$3\frac{1}{4}$
Nutria, plucked	25	$3\frac{1}{4}$
Opossum, natural	37	3
Opossum, dyed	20	3
Opossum, Australian	40	$3\frac{1}{2}$
Otter, land	100	$4\frac{1}{2}$

Otter, sea	100	4½
Persian lamb	65	3¼
Pony, Russian	35	3½
Rabbit	05	2¼
Raccoon, natural	65	2¼
Raccoon, dyed	50	2½
Sable	60	2½
Sable, blended	45	2½
Seal, fur	80	3½
Seal, fur dyed	70	3⅛
Skunk, tipped	50	2⅞
Squirrel, grey	20–25	1¾
Wolf, natural	50	6½
Wolverine	100	7

In estimating the value of a fur, many factors have to be considered. There is no one standard by which the skins are judged, each kind of fur having its own criterion. However, the general points by which raw furs are graded are, color, size, origin, quality and quantity of hair, condition of leather, date or season of trapping, methods of handling, etc. Beaver, for example, is graded as large, medium, small and cubs. Red foxes, first, into Alaska, Labrador, and Nova Scotia, and then these divisions are classed as large, medium and small. Skunks are graded according to the amount of white on the skin, the less white, the more valuable the fur.

The qualities which make a fur desired depend first of all on the nature of the fur itself. Pretty color, luster, thickness, softness, length, uniformity and regular fall of the hair are the chief points to be considered. While the leather part of the fur is of secondary importance in the evaluation of a fur, it must possess strength, lightness of weight, and when properly dressed, should be supple and have a certain firmness or 'feel.' The abundance or scarcity of a fur-bearing animal also determines the value of the fur. Furs which are always comparatively rare, such as silver fox, Russian sable, chinchilla, etc., are always highly prized. In this connection, circumstances which tend to decrease the number of available pelts of any particular animal, such as pestilences, gradual extermination due to excessive trapping, prevention of trapping, by protective laws, also affect the value of a fur. A third factor which has an influence on the value of furs, is the prevailing style or fashion. Many kinds of furs which are both beautiful and rare, such as Russian sable or chinchilla, are practically unaffected by the whims of fashion. But a fur of ordinary value may at times become so popular, that the demand for it will cause its price to be greatly increased. Similarly, a fur which has enjoyed a considerable vogue, may pass out of demand for a time and consequently depreciate in value.

A detailed description of the various furs used in commerce is not within the scope of this work, because such an account rightly belongs in a book on zoölogy. However, it is desirable that the reader who is interested in the dressing and dyeing of furs should have at least a passing acquaintance with the chief furs used in commerce, together with such of their individual characteristics as are of importance. The figures given are for the average dressed skin.

Astrachan, see Lambs.

Badger.—2 × 1 ft. This is one of the few animals whose fur is darker on the belly than on the back. The American sorts have coarse, thick under-hair of a pale fawn or stone color, with a growth of longer black and white hairs 3–4 inches long. The Japanese varieties are usually dyed for imitation skunk. The American kind is also dyed occasionally but is mostly used natural. Badger hair is very extensively used for 'pointing.'

Bear, Black.—6 × 3 ft. Has fine, dark brown under-hair, with bright, flowing black top-hair 4 inches long. The fur of cubs is nearly as long, although the skins are much smaller, and the hair is finer, softer, and lighter-pelted. The best skins are from Canada.

Bear, Brown.—6 × 3 ft. Similar to the Black Bear, but more limited in number. The color ranges from a light yellow to a rich dark brown. The best and most valuable sorts come from the Hudson Bay territory, inferior skins coming from Europe and Asia.

Bear, White.—10 × 5 ft. This is the largest of the bears. The hair is short and close except on the flanks, while the color ranges from white to yellow. The best skins come from Greenland, the whitest being the most valuable.

Beaver.—3 × 2 ft. This is the largest of the rodents, and is very widely used; formerly to a great extent in the hat trade. The under-hair is close and of a bluish-brown hue, and nearly an inch deep. The over-hair is coarse, bright black or reddish-brown in color, and is usually plucked out, as the under-hair is the attractive part of the fur. The darkest skins are the most valuable. Formerly beaver was used to dye in imitation of seal, but more suitable furs are now used.

Broadtail, see Lambs.

Caracul, see Lambs.

Cat, Civet.—9 × 4½ inches, with short, thick and dark under-hair, and silky, black top-hair with irregular white markings. It is similar to the skunk, but is lighter, softer, less full, and has no disagreeable odor.

Cat, House.—18 × 9 inches. Is mostly black and dark brown, the best skins coming from Holland. The hair is weak, coming out with the friction of wear. In the trade, the black variety is known as genet.

Chinchilla.—12 × 7 inches. This is one of the rarest and most beautiful furs. It comes from Bolivia and Peru, where, due to the uncontrolled trapping of the animal, it is becoming scarce, and this compelled the governments to enact laws prohibiting the taking of chinchilla for a certain period. The fur is of a delicate blue-grey, with black shadings, the fur being 1–1¼ inches deep. Unfortunately, the skin is quite perishable.

Chinchilla, La Plata.—9 × 4 inches. Incorrectly called "bastard chinchilla" in the trade. It is a similar species to the Bolivian chinchilla, but due to the lower altitude and warmer climate of its habitat, is smaller, with shorter and less pretty hair, the color of the under-hair being darker, and of the top-hair less pure. It is quite as undurable as true chinchilla.

Chinchillone.—13 × 8 inches. Is also from South America. The fur is longer, weaker, poorer and yellower than real chinchilla, but the skins are often dyed in shades closely resembling the natural chinchilla.

Ermine.—12 × 2½ inches. The under-hair is short and even, with the top-hair slightly longer. The leather is light, close in texture, and quite durable. In mid-winter the color is pure white, except the tip of the tail, which is usually quite black. The best skins are from Siberia.

Fisher.—30 × 12 inches, with tail 12–18 inches long. It is the largest of the marten family. The under-hair is deep, and of a dark shade, with a fine dark, glossy and strong top-hair, 2 or more inches long. The best skins are from Canada. The fur is something like a dark silky raccoon, while the tail, which is very highly prized, is almost black.

Fitch.—12 × 3 inches. It is of the marten species, and its common name is polecat. The under-hair is yellow and ⅓ of an inch deep. The top-hair is black, 1½–1¾ inches long, very fine and open in growth, and not so close as the martens. The largest and best skins are from Denmark, Holland and Germany. The Russian skins are smaller, silkier, and are usually dyed as a substitute for sable.

Fox, Blue.—24 × 8 inches. The under-hair is thick and long, while the top-hair is fine and not so plentiful as in other foxes. It is found in Alaska, Hudson Bay Territory, Greenland and Archangel. Although called blue, the color is really of a slaty or drab shade. The skins from Archangel are more silky and of a smoky bluish color, and being scarce are most valuable. The white foxes which are dyed a smoky blue are brilliant and quite unlike the browner shades of the blue-fox.

Fox, Cross.—20 × 7 inches. The skins generally have a pale yellow or orange tone, with some silver points, and a darkish cross marking on the shoulders, on account of which the animal got its name. Some are very similar to the pale red foxes of Northwest America. The darkest and best skins are from Labrador and Hudson Bay, those from lower latitudes being inferior.

Fox, Grey.—27 × 10 inches. Has a close dark drab under-hair, with coarse regular, yellowish, grizzly-grey top-hair. The majority of the skins come from Virginia and southwestern U. S. A. Those from the west are larger and brighter-toned.

Fox, Kit.—20 × 6 inches. The under-hair is short and soft, as is also the top-hair, which is a very pale grey mixed with some yellowish-white hairs. It is the smallest of the foxes, and is found in Canada and northern United States.

Fox, Red.—24 × 8 inches, although some kinds are larger. The under-hair is long and soft, and the top-hair is plentiful and strong. The colors range from pale yellow to a dark red, some being very brilliant. It is widely found in northern America, China, Japan, and Australia. The Kamchatka foxes are exceptionally fine and rich in quality. Farther north, near the open sea, the fur is coarse. The skins have an extensive use, both natural and dyed. They are dyed black in imitation of the black fox, or these when pointed with badger or other white hair to imitate the silver fox.

Fox, Silver.—30 × 10 inches. The under-hair is close and fine, and the top-hair, which is black to silvery, is 3 inches long. The fur on the neck usually runs almost black, and in some cases the black extends over half the length of the skin. When all black, it is a natural black fox, and is exceedingly rare and high-priced. The silver fox is very valuable, the finest wild skins coming from Labrador. The tail is always tipped white. The majority of the silver fox pelts that reach the market today are bred on ranches in Canada and the United States.

Fox, White.—20 × 7 inches. It is usually small and inhabits the extreme northern sections of Hudson Bay, Labrador, Greenland, and Siberia. The Canadian are silky-haired and inclined to a creamy color, while the Siberian are whiter and more woolly. The under-hair is generally of a bluish-grey tone, but the top-hair in winter is usually full enough to hide such a variation. Those skins which have under-hair that is quite white are rare and much more expensive than the others. In summer specimens of these species have slightly discolored coats, the shades resembling those of the blue fox. The skins which are not perfectly white are bleached, or if they cannot be bleached sufficiently white they are dyed various shades of smoke color, blue-greys and also imitation blue fox.

Goat.—The size varies greatly. The European, Arabian and East Indian varieties are used mainly for leather and wool. Many from Russia are dyed black for rugs. The hair is brittle, with poor under-hair, and is not durable. The Chinese export many skins in grey, black and white, made into rugs of two skins each. Frequently the skins are dyed black or brown in imitation of bear.

Hamster.—8 × 3½ inches. A destructive rodent found largely in Russia and Germany. The fur is very flat and poor, of a yellowish-brown color, with a little marking of black. On account of its lightness it is used for linings; occasionally it is dyed.

Hare.—24 × 9 inches. The common hare of Europe is used mostly for the hatters' trade. The white hares of Russia, Siberia, and other northern regions are the ones mainly used for furs. It is whitest in mid-winter, and the fur on the flanks is longer than that on the back. The hair is brittle and not durable, and the leather is quite as bad. Yet the skins are used to dye imitations of more than a dozen different furs. The North American hares are also dyed black and brown.

Kangaroo.—The sizes vary greatly, the larger kinds being generally used for making leather. The sorts used for fur are, blue kangaroo, bush kangaroo, wallaroo, rock wallaby, swamp wallaby, and short-tailed wallaby. Many of the swamp wallabies are dyed imitation skunk, and look quite attractive. The colors are generally yellowish or brown, some in the swamp variety being dark brown. The skins are quite strong. The rock wallabies are soft and woolly, and often have a bluish tone. They are used for rugs.

Kolinsky.—12 × 2½ inches. It is of the marten family. The under-hair is short and rather weak, but regular, as is also the top-hair. The color is usually a uniform yellow. They are generally dyed in imitation of other members of the marten family. It is very light in weight, and the best skins are obtained from Siberia. The tails are used for artists' "sable" brushes.

Lambs.—Those of commercial interest are from South Russia, Persia, and Afghanistan, and include Persian Lamb, Broadtail, Astrachan, Shiraz, Bokhara, Caracul, and Krimmers.

The *Persians* are 18 × 9 inches, and are the finest and best. When properly dressed and dyed they should have regular, close, bright curls, varying from small to very large and if of equal size, regularity, tightness and brightness, their value is inestimable.

All the above lambs, except krimmer, are naturally a rusty black or brown, and are in most cases dyed a jet black. Luster cannot be imparted where naturally lacking.

Broadtails, 10 × 5 inches, are the young of the Persians, killed before the wool has had time to develop beyond the flat wavy state. They are naturally of exceedingly light weight, and when of an even pattern possessing a lustrous sheen are costly. The pelt, however, is too delicate to resist hard wear.

Astrachan, *Shiraz*, and *Bokhara* lambs, 22 × 9 inches, are of a coarser and looser curl. Caracul lambs are the very young of the astrachan, and the finest skins are almost as effective as the broadtails, although not so fine in texture.

Krimmers, 24 × 10 inches are grey lambs obtained from Crimea. They are of a similar nature to the caraculs, but looser in curl, and ranging in color from a very light to a dark grey, the best being pale bluish-greys.

Slink lambs come from South America and China. The South American are very small, and generally those are still-born. They have a particularly thin pelt, with very close wool of minute curls. The Chinese sorts are much larger.

Leopard.—3 × 6 feet long. There are several kinds, the chief being the snow leopard or ounce, Chinese, Bengal, Persian, East Indian, and African. The first variety inhabits the Himalayas, and has a deep, soft fur, quite long as compared with the Bengal sort. The colors are pale orange and white with dark markings. The Chinese are of a medium orange-brown color and full in fur. The East Indian are less full and not so dark; the Bengal are dark and medium in color with short, hard hair. The African are small, with pale lemon-colored ground, and very closely marked with black spots.

Lynx.—45 × 20 inches. The under-hair is thinner than in the fox, but the top-hair is fine, silky and flowing, 4 inches long, of a pale grey, slightly mottled with fine streaks and dark spots. The fur on the flank is longer, and white, with very pronounced markings of dark spots, and this part of the skin is generally worked separately. Skins with a bluish tone are more valuable than those with a sandy or reddish hue. The lynx inhabits North America as far south as California. The best skins come from Hudson Bay, and also Sweden. They are generally dyed black or brown, similar to dyed fox.

Marmot.—18 × 12 inches. A rodent found largely in the south of Germany. The fur is yellowish-brown, rather harsh and brittle, and without under-hair. Also found in North America, China, and the best skins come from Russia. It is dyed brown in imitation of mink or sable, the stripes usually being put on in the completed garment.

Marten, Baum.—16 × 5 inches. Also called Pine Marten, and is found in the woods and mountains of Russia, Norway, Germany and Switzerland. It has a thick under-hair with strong top-hair, and ranges from a pale to a dark bluish-brown. The best are from Norway, are very durable and of good appearance, and a good substitute for the American sable.

Marten, Japanese.—16 × 5 inches. It is of a woolly nature with rather coarse top-hair, and quite yellow in color. It is dyed, but it is not an attractive fur, lacking a silky, bright and fresh appearance.

Marten, Stone.—Size and quality similar to the baum marten. The color of the under-hair is stony white, and the top-hair is a very dark brown, almost black. Skins of a pale bluish tone are used natural, while less clear colored ones are dyed, usually in Russian sable shades. They are found in Russia, Bosnia, Turkey, Greece, Germany, and France, the best coming from Bosnia and France.

Mink.—16 × 5 inches. Is of the amphibious class, and is found throughout North America, as well as in Russia, China and Japan. The under-hair is short, close and even, as is also the top-hair, which is very strong. The best skins are very dark, and come from Nova Scotia. In the central states the color is a good brown, but in the northwest and southwest, the fur is coarse and pale. It is very durable and an economic substitute for sable. The Russian species is dark, but poor and flat in quality, and the Chinese and Japanese sorts are so pale that they are always dyed.

Mole.—3½ × 2½ inches. Is plentiful in the British Isles and Europe, and is much in demand on account of its velvety fur of a pretty bluish shade. Although the skins are comparatively cheap, the cost of dressing is high on account of the considerable amount of labor involved. The pelt is very light in weight, but does not resist well the friction of wear.

Monkey, Black.—18 × 10 inches. The species usually found on the west coast of Africa, is the one of interest to the fur trade. The hair is very long, very black and bright, with no under-hair, and the white pelt is very noticeable by contrast.

Muskrat, Brown, Black, Russian.—12 × 8 inches. A very prolific rodent of the amphibious class, obtained in Canada and the United States. It has a fairly thick and even brownish under-hair, and a rather strong, dark top-hair of medium density. It is a durable and not too heavy fur. It is used natural, but recently the plucked, sheared and dyed skins have found a very extensive use as Hudson seal, an imitation of real seal. The so-called black variety of muskrat is found in New Jersey and Delaware, but only in comparatively small numbers. The Russian is also very small and limited in numbers. It is of a pretty silvery-blue shade with even under-hair, with very little silky top-hair, and silvery-white sides, presenting altogether a marked effect.

Nutria.—20 × 12 inches. Is a rodent about half the size of the beaver, and when plucked, has only about half the depth of fur, which is not so close. It is often dyed a seal color, but its woolly nature renders it less effective than the dyed muskrat. The skins are obtained from northern South America.

Opossum, American.—18 × 10 inches. Is a marsupial, the only one of its class found outside of Australia. The under-hair is of a very close frizzy nature, and nearly white, with long bluish-grey top-hair mixed with some black. It is found in central sections of the United States, and is frequently dyed imitation skunk.

Opossum, Australian.—16 × 8 inches. Is of a totally different nature from the American. Although it has fur-hair and top-hair, the latter is sparse and fine, so that the fur coat may be considered one of close even under-hair. The color varies according to the district of origin, from blue-grey to yellow with reddish tones.

Those from near Sidney are a light clear blue, while those from Victoria are a dark iron-grey, and stronger in the fur-hair. The most pleasing shade of grey comes from Adelaide. The reddest are the cheapest. The ring-tailed opossum, 7 × 4 inches, has a very short, close and dark grey under-fur, some almost black, but the skins are not used extensively. The Tasmanian opossum, grey and black, 20 × 10 inches, is of a similar description, but larger, darker, and stronger in the under-hair.

Otter, River.—The size varies considerably, as does also the length of the fur, according to the origin. It is found in greatest numbers in the coldest northern regions, and with the best under-hair, the top-hair being unimportant, as it is plucked out. Most of the best river otters come from Canada and the United States, and average 36 × 18 inches. The skins from Germany and China are smaller and shorter furred. The colors of the under-hair vary from very dark brown to almost yellow. Both the fur and the leather are extremely strong, and many skins are dyed imitation seal after plucking.

Otter, Sea.—50 × 25 inches. Is one of the most beautiful of furs. The under-hair is of a rich, dense, silky nature, with short and soft top-hair, which is not plucked. The colors range from a pale grey-brown to a rich black, and many skins have a sprinkling of white or silver-white hairs. The blacker the under-hair, and the more regular the silver points, the more valuable is the skin.

Pony, Russian.—This is a comparatively cheap, but very serviceable fur, and possesses some very desirable qualities. It has a thin leather, but is also scantily haired. Young pelts have a design on them somewhat similar to broadtail lambs, or moire astrachans, but this design is lost to a considerable degree by dyeing the furs. The hair, which is very glossy, is generally dyed black, although the natural pelts are also worn extensively.

Rabbit.—10 × 16 inches. The fur is thick and fine, but the pelt is very weak. It is a native of central Europe, Asia, North and South America, New Zealand and Australia. The color ranges from white to black. France, Belgium and Australia are the greatest producers of rabbits suitable for dyeing black, the so-called French seal, for which they are mostly used. At the present time the dyeing of rabbits constitutes a considerable percentage of the total fur-dyeing operations in this country. The most varied shades are produced on rabbit, and it probably is the basis of the greatest number of dyed imitations of better furs. In addition to the French seal, or sealine, rabbit is dyed in imitation of beaver, mole, etc.

Raccoon.—20 × 12 inches. Varies considerably in size, quality and color of the fur, according to the part of North America in which it is found. The under-hair is 1–1½ inches deep, pale brown, with long top-hair of a dark and silvery-grey mixture of a grizzly type, the best having a bluish tone, and the cheapest a yellowish or reddish-brown. The best skins come from the northern part of the United States. The skins have a wide use natural, but are also dyed dark blue, or imitation skunk, the latter being a very effective and attractive substitute, and extensively used. Sometimes the skins are plucked, and if the under-hair is good, the effect is similar to a beaver.

Sable, American and Canadian.—17 × 5 inches. The skins are sold in the trade as martens, but since many of the skins are of a very dark color, and almost as silky as Russian sable, they have come to be known as sable. The prevailing color is a medium brown, while many are quite yellow. These pale skins have been dyed so well that they can cheaply substitute Russian sable. The finest skins are from the Eskimo Bay and Hudson Bay districts, the poorest from Alaska.

Sable, Russian.—15 × 5 inches. Belongs to a species of marten similar to the European and American, but much more silky in the texture of the fur. The under-hair is close, fine and very soft, the top-hair is regular, fine and flowing, and silky, ranging from 1½ to 2½ inches in depth. In color they vary from a pale stony or yellowish shade to a rich, almost black, dark brown, with a bluish tone. The leather is exceedingly close and fine in texture, very light in weight, and very durable. The Yakutsk, Okhotsk, and Kamchatka sorts are good, the last being the largest and fullest-furred, but of less color density than the others. The most valuable, are the darkest from Yakutsk in Siberia, particularly those having silvery hairs evenly distributed over the skin, but these furs are very rare.

The Amur skins are paler, but often of a pretty, bluish tone, with many interspersed silvery hairs. The fur is not so close or deep, but is very effective nevertheless. The paler skins from all districts are now tipped, the tips of the hair being stained dark, the fastest dyes being used, and only an expert can detect them as differing from the natural shades.

Seal, Fur.—The sizes range from 24 × 15 inches to 15 × 25 inches, the width being the widest part of the skin after dressing. The most useful skins are the pups 42 inches long, the quality being very good and uniform. The largest skins, known as wigs, and ranging up to 8 feet in length, are uneven and weak in the fur. The supply of the best sort is chiefly from the northern Pacific, Pribilof Islands, Alaska, northwest coast of America, Aleutian Islands, and Japan. Other kinds are taken from the south Pacific regions. The dressing and dyeing of seal takes longer than for any other fur, but when finished, it has a fine, rich effect, and is very durable.

Seal, Hair.—This is chiefly used for its oil and leather, and not for its fur. It has coarse, rigid hair, and no under-hair.

Skunk, or "Black Marten."—15 × 8 inches. The under-hair is full, and fairly close, with glossy, flowing top-hair about 2½ inches long. The majority of the skins have two stripes of white hair extending the whole length of the skin. These were formerly cut out, but more recently are dyed the same color as the rest of the skin. They are widely found in North and South America. The best are from Ohio and New York. The skunk is naturally the blackest fur, is silky and very durable.

Squirrel.—10 × 5 inches. This size refers to the Russian and Siberian types, which are practically the only kind imported for fur, other species having too poor a fur to be of great commercial interest. The back of the Russian squirrel has an even, close fur, varying from a clear bluish-grey to a reddish-brown, the bellies in the former being of a flat quality and white, in the latter, yellowish. The backs are worked up separately from the bellies. The pelts, though light in weight, are tough and durable. The tails are dark and very small, and considerably used.

Tiger.—The size varies, the largest measuring about 10 feet from the nose to the root of the tail. It is found throughout India, Turkestan, China, Mongolia, and the East Indies. Coats of the Bengal variety are short and of a dark orange-brown with black stripes. Those from other parts of India are similar-colored, but longer in hair, while those from the north and China are not only large in size, but have very long soft hair of a delicate orange-brown, with very white flanks, and marked generally, with the blackest of stripes.

Wolf.—50 × 25 inches. Is closely allied to the dog family, and very widely distributed over the world. The best are the full-furred skins of a very pale bluish-grey with fine, flowing black top-hair, from the Hudson Bay district. Those from the United States and Asia are harsher and browner. The Siberian is smaller than the North American, and the Russian still smaller. A large number of prairie-dogs, or dog-wolves, is also used for cheaper furs.

Wolverine.—16 × 18 inches. Is a native of America, Siberia, Russia, and Scandinavia, and is of the general nature of the bear. The under-hair is full and thick, with strong, bright top-hair about 2½ inches long. The color is of two or three different shades of brown on one skin, the center being dark, and presenting the general appearance of an oval saddle, bordered with a rather pale shade of brown, and merging to a darker shade towards the flanks. This peculiar character stamps it as a distinguished fur. It is expensive, and quite valuable on account of its excellent qualities.

Wombat, Koala or Australian Bear.—20 × 12 inches. It has a light grey or brown, close, thick under-hair ½ inch deep, and no top-hair, with a rather thick, spongy pelt. It is cheap, and well suited for rough wear.

CHAPTER II
STRUCTURE OF FUR

Fur is made up of two main components, the hair and the skin, and each of these has a very complex structure.

In the living animal the skin serves as a protective covering, and also constitutes an organ of secretion and of feeling; consequently it is of a highly complicated nature. The skin of all fur-bearing animals is essentially the same in structure, although varying considerably as to thickness and texture. It consists of two principal layers, which are entirely different in structure and purpose, and correspondingly different in both physical and chemical respects: the epidermis, epithelium or cuticle, which is the outer layer, and the dermis or corium, which is the true skin. (Fig. 1A).

The epidermis is very thin as compared with the corium. Its outer layer consists of a tissue of cells, somewhat analogous to the horny matter of nails and hair. The inner surface, called the 'rete malpighi,' rests on the true skin, and is a soft, mucous layer of cells. These cells are spherical when first formed, but as they approach the surface become flattened, and dry up, forming the horny outer layer of the epidermis, which is constantly throwing off the dead scales, and which is constantly being renewed from below. It is from this inner layer of the epidermis that the hair, the sweat-glands, and the fat-glands are developed.

The corium, or true skin, consists essentially of white, interlacing fibres of the kind known as connective tissue. These fibres are themselves made up of extremely fine smaller fibres, or fibrils, cemented together by a substance of a somewhat different nature from the fibres, the coriin. Towards the center of the skin, the texture of the interweaving fibres is looser, becoming much more compact at the surface just beneath the inner layer of the epidermis. This part of the corium is so exceedingly close that the fibrils are scarcely recognizable. It is in this part that the fat-glands are situated, while the hair-roots and sweat-glands pass through it into the looser texture of the corium. The surface next to the flesh is also closer in structure than the middle portions of the skin, and has somewhat of a membranous character due to the fibres running almost parallel to the surface of the skin. The skin is joined to the body proper by a network of connective tissue, frequently full of fat-cells. This layer, together with portions of the flesh which may adhere to it, is removed by the process called 'fleshing,' and this side of the skin is known as the flesh side. The corium also contains a small proportion of yellow fibres, known as 'elastic fibres,' which differ physically and chemically from the rest of the skin substance.

During the course of the development of the embryo animal, a small group of cells forms like a bulb on the inner side of the epidermis, above a knot of very fine blood-vessels in the corium. This group of cells grows downward into the true skin, and the hair-root which is formed within it, surrounds the capillary blood-vessels, drawing nourishment from them, and thus forming the papilla. (Fig. 1A). Smaller projections also form on the bulb, and the fat-glands are gradually developed. The sweat-glands are formed in a manner similar to the development of hair.

The individual hair fibre is quite as complicated in structure as the skin, and is made up of four distinct parts. (Fig. 1B).

The medulla, or pith, is the innermost portion of the hair, and is composed of many shrunken cells, often connected by a network which may fill the medullary column partially or wholly.

Surrounding the medulla is the cortex, which is made up of spindle-shaped cells fused into a horny, almost homogeneous, transparent mass, and forming a large proportion of the hair shaft.

In the majority of the fur-bearing animals, there is distributed within and among the cells of the cortex a pigment in the form of granules or minute particles, arranged in the different hairs in fairly definite and characteristic patterns. It is to these pigment granules that the color of the hair is due primarily. In some cases the coloring matter of the shaft is uniformly diffused and not granular.

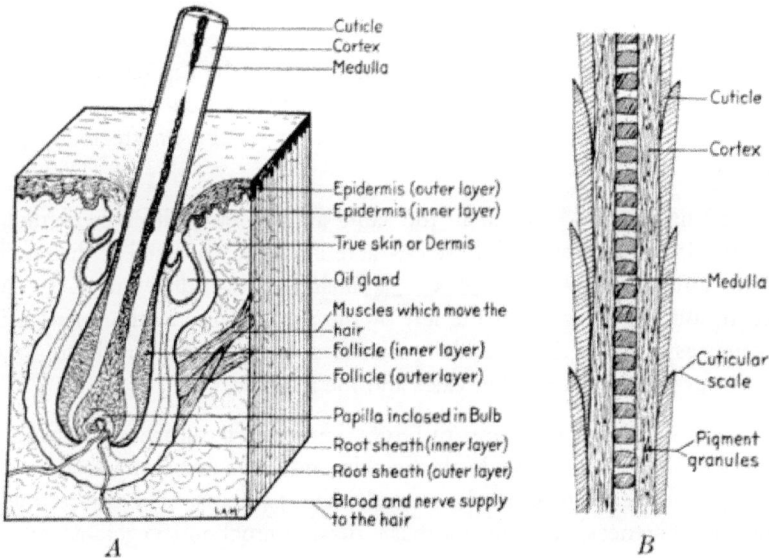

Fig. 1

A. STRUCTURE OF SKIN.   *B*. STRUCTURE OF HAIR.

The outermost coat of the hair, or cuticle, is composed of thin, colorless, transparent scales of varying forms and sizes, and arranged in series like the shingles of a roof. It is on these scales that the lustre or gloss of the hair depends. Since lustre is due to the unbroken reflection of light from the surface of the hair, the smoother the surface, the glossier it will appear. When the scales of the cuticle are irregular and uneven, the surface of the hair will not be uniform and smooth, and the light reflected from it will be broken and scattered, and consequently the hair will not possess a high degree of lustre. As a rule, the stiff, straight hairs have the most regular and uniform arrangement of the scales of the cuticle, and hence are the smoothest and glossiest.

Fur hairs are in general either circular or elliptical in cross-section, those which are circular being straight or only slightly curved, while those which are elliptical in cross-section are curly like the hair of the various kinds of lambs.

Most fur-bearing animals have two different kinds of hair on their bodies. Nearest to the skin is a coat of short, thick, soft and fine hair, usually of a woolly nature, and called the under-hair, under-wool, or fur-hair. Overlying the fur-hair is a protective layer of hair, longer and coarser than the under-hair, and usually straight, hard, smooth and glossy. This is called the top-hair, over-hair, guard-hair or protective hair. In some furs, the top-hair constitutes one of the chief elements of their beauty, while in others, the top-hairs are removed, so as better to display the attractive features of the under-hair. The roots of the top-hair are generally deeper in the skin than those of the fur-hair, and in some instances where the top-hair is removed, as in the seal, the roots are destroyed by the action of chemicals applied to the skin side, the roots of the fur-hair being wholly unaffected by this treatment.

The fur-hair and the top-hair in the same animal have different medullary and cuticular structures, and these characteristics may be used to distinguish the two kinds of hair. Figs. 2A and B illustrate these differences. In each case, the two large hairs on the left of the illustration are the guard-hairs, showing respectively the cuticular scales and the medulla. On the right are the two fur-hairs showing the scales and the medulla.

Although composed of many different kinds of tissues, and varying so greatly in physical structure, both the skin and the hair belong to the same class of chemical compounds, namely the proteins. These are highly complex substances, forming the basis of all animal and vegetable tissues.

There are many different kinds of proteins, varying somewhat in their constitutions, but all show, on analysis the following approximate composition of chemical elements:

Carbon	50–55%
Hydrogen	6.5–7.3%
Nitrogen	15–17.6%
Oxygen	19–24%
Sulphur	0.3–5%

The principal kinds of proteins found in the various fur structures are albumins, keratin, collagen, and mucines. Albumins, of which the white of egg is the most familiar variety, occurs to some extent in the corium as serum in the blood-vessels, and also as the liquid filling the connective tissues, known as the lymph. They are soluble in cold water, but when heated to about 70° C., they coagulate and are then insoluble. Concentrated mineral acids and strong alcohol will also effect coagulation.

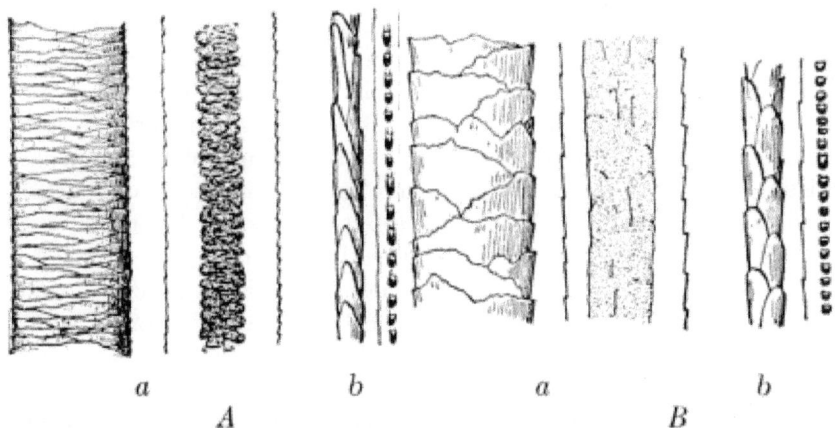

Fig. 2

A. HAIR OF EUROPEAN BEAVER.   *B*. HAIR OF SKUNK.

a. TOP-HAIR. *b*. UNDER-HAIR.   *a*. TOP-HAIR. *b*. UNDER-HAIR.

Keratin is the chief substance of which all horny parts of the animal body are composed, such as the hair, nails or hoofs. It is the principal constituent of the hair, the epidermis, and the walls of the cells of the inner layer of the epidermis, or the 'rete malpighi.' Keratin is particularly rich in sulphur, and is quite insoluble in cold water. Caustic alkalies attack keratin-containing parts.

The collagens are the principal proteins of the skin, forming largely the substance of the connective tissue fibres, and consequently the framework of the skin. They are insoluble in cold water, dilute acids and salt solutions, and are only very slowly attacked by dilute alkalies. Dilute acids and alkalies cause collagen to swell; concentrated acids, vegetable tanning materials, basic chrome or iron salts cause it to shrink. By boiling with water, dilute acids or dilute alkalies, collagen is split up into gelatin or glutin.

The mucines of the skin, intercellular material or coriin, are soluble in dilute acids, in dilute solutions of alkalies and of alkaline earths such as lime, and in 10% salt solution, but insoluble in water, and in salt solutions of greater or less concentration than 10%. On drying the skin, the mucines cement the connective tissue fibres, causing the skin to become stiff, horny and translucent. The mucines are also constituents of the cells of the 'rete malpighi.' The solubility of the mucines in dilute solutions of alkalies and of alkaline earths causes the epidermis to be loosened from the corium, when the skins are treated with such solutions for some time.

When raw skins are boiled with water, the greater part goes into solution, the residue consisting chiefly of the keratins of the hair and epidermis cells. On cooling, the solution solidifies to a jelly of gelatine. It combines with both acids and alkalies. A property of the skin which is of importance in the tanning operation of fur-dressing, and a quality which also characterizes gelatine, is the capacity to absorb liquids and swell up, without changing chemically. Raw pelts swell up easily in pure cold water, but much more easily in solutions of dilute acids or dilute alkalies, only a little of the skin material being dissolved. In stronger solutions, the skins swell up less, while more of the skin substance dissolves, and by prolonged action of strong acids or alkalies, an almost complete solution of the skin is obtained, without, however, any of the material decomposing. With very strong alkalies or acids, the skin substance is broken up into simpler compounds, such as various amines and ammonia. The swelling action of acids or of alkalies increases with the increase in concentration of the acid or alkali, but only up to a certain point, after which further increase in the strength of the acid or alkaline solution causes a reduction in the swelling, and even produces shrinkage. In the presence of neutral salts, like common table salt, sodium chloride, the swelling action of acids, is reduced, but the action of alkalies remains practically unaffected.

When treated with the various chemicals, fur hair acts in a manner quite similar to wool. If it be remembered that certain classes of furs are derived from animals of the sheep family, such furs as Persian lamb, krimmer, etc., it becomes apparent why chemicals should affect furs in nearly the same way as wool. The great majority of furs differ from those of the sheep family, in possessing much greater resistance to the action of chemicals. The range is a wide one however, and no exact criterion can be adopted. As a general rule, the reactions are most marked with fur-hair of a woolly nature, so this may be taken as a standard of reference.

Acids have relatively little action on the hair, when applied in dilute solutions. The scales of the cuticle or epithelium are somewhat opened, the fibre becoming slightly roughened thereby. Even at high temperatures, the hair is quite resistant to the action of dilute acids. Concentrated acids destroy the hair with the liberation or formation of ammonia, hydrogen sulphide, and various amino acids. When treated with dilute acids, the hair, especially if it is of a very woolly nature, retains considerable quantities of acid, this phenomenon being probably due to the fixation of the acid by the basic groups in the hair. Nitric acid produces a yellow coloration when applied in dilute solution for a short time. Sulphurous acid, the acid formed by the burning of sulphur, has a bleaching action on the hair.

Alkalies attack the hair, even in dilute solutions, and by longer action complete decomposition sets in, with formation of ammonia and amino-acids. Ammonium carbonate, soap, and borax are practically harmless in their effect on the hair. Sodium and potassium carbonates roughen the hair on prolonged action, even in dilute solutions. Calcium hydroxide on continued action removes sulphur from the hair, causing it to become brittle.

Salts of alkalies and alkaline earths do not affect the hair at all. Salts of the heavy metals on the other hand, are absorbed in appreciable quantities. From a dilute solution of alum, aluminum hydroxide is absorbed by the hair, the potassium sulphate remaining in solution. Similarly with copper, iron, and chromium salts, the metal oxides are fixed by the fibre.

CHAPTER III
FUR DRESSING
INTRODUCTORY AND HISTORICAL

Fur dressing has a twofold purpose. First of all, the putrefactive processes must be permanently stopped, so that the skin may be preserved as such, or worked up as some fur garment, without danger of decomposition. Having taken measures to assure the endurance or relative permanency of the pelt, the prime consideration is, of course, the appearance of the hair. The hair must be so treated that all its inherent beauty is brought out to the fullest extent. It must be made clean and soft, and all the natural gloss must be preserved, and if possible, enhanced. The appearance of the leather is relatively unimportant, since it is not seen after the furs are made into garments. There are, however, certain qualities which it is essential for the leather to possess after being dressed, and these are, softness, lightness of weight, elasticity or stretch, and a certain firmness or 'feel.' In other words the important considerations in fur dressing are the employment of means, and the exercise of care to preserve or even improve those characteristics of the pelt which make it valuable.

The dressing of furs has many features in common with the manufacture of leather, which is a kindred art. But whereas in fur dressing the prime consideration is the appearance of the hair, and the leather is of secondary importance, in the production of leather, the hair plays no part at all, since it is entirely removed from the pelt. The fundamental points of resemblance between leather manufacture and fur dressing are in those processes and operations which are concerned with the preservation of the leather, and rendering it in the proper condition for use.

Both leather dressing and fur dressing have an origin which may be regarded as identical, and which dates back to the haziest periods of antiquity. In the course of satisfying his needs, primitive man killed the animals about him, and thus obtained his food. The killed animal also furnished a skin, which after undergoing certain manipulations and other treatments, could serve as a protective covering, ornament, or defensive weapon. Since the skin in its natural state was hardly fit for use because of its easy tendency to putrefaction, it is evident that man had to find some means of preventing this decay in a more or less permanent fashion, and moreover had to treat the skin so that it would be suitable for use, by rendering it soft and flexible. The discovery of means to accomplish these purposes was probably one of the first great steps forward on the path of progress and civilization.

There are evidences of the use of animal skins in the earliest periods of antiquity, in fact it is a usage which may be literally regarded as "old as the hills." One of the earliest written records of the employment of the skins of animals as garments, is in the Old Testament, where it states, "Unto Adam and to his wife did the Lord God make coats of skins, and clothed them." Numerous other biblical references indicate the use of animal skins for various purposes, sometimes prepared as leather, with the hair removed. Among the Egyptians tanning seems to have been a common occupation. The particularly attractive skins, like those of the leopard or panther, were especially prized, and were made up as furs for ornamental wear, rugs and decorations. The less valuable skins were unhaired and made into leather. Although the tanning or leather-producing processes of the Egyptians are quite unknown, numerous figures engraved in stone afford an indication to some of the manipulatory operations, such as soaking the skins, fleshing, softening with stones, stretching over a three-legged wooden "horse," etc. Many articles, made of leather, have been found in the various Egyptian sarcophagi, and all are in a splendid state of preservation, after forty centuries, thereby indicating a very efficient method of dressing animal skins. Likewise, the presence in the museums of various articles, leather and fur, of Assyrian, Phoenician and Persian origin, tends to show that these peoples also possessed a considerable degree of proficiency in tanning. Frequent references in the Greek literature show that leopard and lion skins were worn as war cloaks, and they undoubtedly were properly made. In the *Iliad* is described an operation for the preparation of skins for use as garments, and the method seems to be a sort of chamois dressing.

The first method of tanning skins was, in all probability, that of rubbing into the skins various fatty materials found close at hand, such as parts of the animal, fat, brains, milk, excrement, etc., such an operation constituting the basis of what is now known as the chamois dressing. One of the reasons for believing that it was the first process to be used by primitive man, is the fact that certain undeveloped tribes and races of the present day still dress skins by it. The American Indians, even to this day prepare skins by rubbing in, on the flesh side, the brains of the animals which furnished the skins. The Eskimos dress skins by rubbing in animal fats or fish-oil, and subsequently softening and stretching the skins with their teeth in place of, or for want of other implements. Usually, however, variously shaped stones or bones of animals are used to obtain the proper degree of softness and flexibility. It is true, too, that some of the skins dressed in this primitive fashion can scarcely be excelled by any dressed with more modern processes and tanning methods.

The next step forward in the preparation of animal skins for use was undoubtedly the utilization of substances found in the earth. Common salt, sodium chloride, was the most universally used substance of mineral origin, just as it is today. Our prehistoric ancestors eventually discovered the preservative action of salt, and applied it to skins. While it was effective, it was not sufficiently permanent, so another mineral, also of very common and wide occurrence was used in combination with the salt, and the result proved quite satisfactory. This second common mineral was alum. The use of alum, which is the basis of numerous tanning processes to this day, seems to have been quite a popular method of ancient times. Artemidorus, a Greco-Roman writer, mentions the use of alum by the Greeks, and the Romans are known to have prepared a soft, flexible leather called aluta (alum leather), by using it. In view of the fact that Egypt had extensive deposits of alum, it is believed that the alum-salt process was employed also by the Egyptians in the preparation of leather. However, the evidence on this point is not conclusive.

One of the most important methods of producing leather, either as such or on furs, was with the aid of certain vegetable extracts, known as the tannins, from which the process of tanning gets its name. The discovery of the value of these materials for converting the decaying raw skin into a leather which could be preserved for an almost indefinite length of time, and which was flexible and soft as desired, was of far-reaching importance. For it is only in very recent times that these tannins have been superseded in part by new tanning substances whose use is simpler and more time-saving. Yet there are unmistakable indications that the tannins were employed for tanning at a period which reaches back to the dawn of history. Although it is scarcely probable that the people who used these materials could have known of the existence or the nature of the particular substances in the vegetable extracts which actually effect the tanning action, experience taught them to employ these plants which possessed the highest content of active ingredients, and which, consequently, were most effective in use. Tychios, of Boetius, a Greek supposed to have lived about 900 B.C. and mentioned in the *Iliad*, is considered the oldest known tanner, and was regarded by Pliny, a Roman writer, as the discoverer of tanning, and of the use of the various vegetable tanning materials. At any rate, the Greeks used the leaves of a so-called tanning-tree, which was probably the sumach. The Egyptians worked with the acacia, while the Romans used as tanning materials the barks of the pine, alder and pomegranate trees, also nut-galls, sumach and acorns. The Romans were quick to employ methods used by the peoples whom they conquered, and it is in this way that they learned the use of many of the plants mentioned, for tanning purposes.

Many other ancient peoples had various processes of tanning, the methods probably differing in each country. Thus the Chinese, Syrians, and much later, the Moors, were each known for proficiency in a certain class of leather tanning. It has been said that in general, even up to modern times, tanning with nut-galls was the characteristic method of the Orient; with oak-tan, that of the Occident, while the use of alum is regarded as the method peculiar to the Saracens.

In prehistoric times and the early centuries of civilization, skins or pelts were prepared for use by the individual, the work usually being done by the housewife and daughters, while the masculine members of the family were engaged in hunting the animals and obtaining the skins. At a later period, when people had advanced to the point where they lived in cities, the preparing or dressing of skins became centered in the hands of a comparatively small number of people, and thus the work took on the aspects of a trade. The workers in fur were at first the same people who made leather out of the skin, for the two kinds of work were very closely associated. During the period of the Roman supremacy, historical records show that the furriers, who did all the work connected with furs, from purchasing the raw skins, dressing them, making them into garments, to selling the latter, were organized into associations together with the leather workers. After the fall of the Roman empire, and throughout the centuries known as the Dark Ages, all traces of the furriers seem to have been lost, but in the beginning of the Renaissance period in the fourteenth and fifteenth centuries, we again find records of the furriers, who were now all members of the furriers' guilds, also in association with the leather workers. As formerly, all the work connected with the production of fur apparel from the raw furs, was done by the master furrier and his apprentices. The methods and the implements used, were essentially the same as in Roman times, and in fact, up to a very recent period there was very little change in either.

With the advent of the great industrial era at the beginning of the nineteenth century, the guild system became ineffective, but the furriers continued their work as heretofore. Up to about the middle of the nineteenth century, the furrier continued to be the only factor of any importance in the fur trade. There was no need for speed in his work, for the demands of the trade were not so urgent. The fact that the dressing of furs often occupied two to four weeks was no deterring factor in his business. However, with the great expansion of the fur trade about this time, it became impossible for the individual furrier to do everything himself, and keep up with the requirements of his customers. Specialization commenced, and establishments were set up solely for fur dressing. The traditional time- and labor-consuming processes were still used, but the efficiency of work on a large scale enabled the fur dressers successfully to fill their orders. But the fur trade continued to grow by leaps and bounds, and very soon the fur dressers were no longer able to meet the demands of the trade. It was then that the science of chemistry came to the aid of the fur dresser, and helped him meet the exigency. By devising dressing processes which were cheap and efficient, and which only required several hours, or at the most one or two days, as compared with as many weeks, the chemist brought the fur dresser out of his dilemma. And with the adoption of mechanical time- and labor-saving devices, the fur dressing industry has made wonderful progress.

CHAPTER IV
FUR DRESSING
PRELIMINARY OPERATIONS

The fur dresser receives the skins in one of two shapes, flat or cased, depending on the manner in which they were removed from the animal. Flat skins, as for example, beaver, are obtained by cutting on the under side of the animal from the root of the tail to the chin, and along the inner side of the legs from the foot to the first cut. The skins are either fastened to boards or attached to wooden hoops slightly larger than the skins, so as to stretch them, and are then carefully dried, avoiding direct sunshine or artificial heat, as it is very easy to overheat the skins and thereby ruin them. The great majority of skins, however, are cased. The pelts are cut on the under side of the tail, and along the hind legs across the body, the skin being then removed by pulling it over the head off the body like a glove, trimming carefully about the ears and nose. The skin is thus obtained inside out, and is drawn over a stretching board or wire stretcher of suitable shape and dimensions, so as to allow the skin to dry without wrinkling. The pelts, after drying in a dry, airy place, are removed from the stretchers and are ready for the market. With some furs, as foxes, the skins are turned hair-side out while still somewhat moist, and then put on the stretcher again till fully dried. In most cases, however, skins are sold flesh-side out. Throughout the various dressing operations cased skins are kept intact, being turned flesh-side out or hair-side out according as the processes are directed to the respective sides. The pelts are only cut open if they have to be dyed, or after the manufacturer receives them, when they have to be worked into manufactured garments.

A distinction which is made by fur dressers and dyers, and also by the fur trade in general, divides furs into those derived from domestic animals, particularly the various kinds of sheep, including also the goat species, and those obtained from other animals by trapping. In fact, at one time, and to a certain extent even to-day, dressers were divided into two groups based on this distinction, one class dealing only with furs obtained from the sheep family, and the other working with other kinds of furs. This differentiation is not a simple arbitrary one, but has a rational justification. As mentioned before, the manner and habit of living of the animal are important factors in determining the nature and constitution of its skin, both leather and hair. The structure of the body being dependent primarily upon the nature of the food absorbed by the animal, it is only natural that herbivorous or vegetable-eating animals such as sheep and goats, should possess fur of a different sort from that of the carnivorous or meat-eating animals, such as the majority of fur-bearers are. It also seems clear that furs differing in their character and constitution should require somewhat different treatments, and accordingly the methods are modified when furs like lambs or goats are dressed. To a great extent, however, the fundamental operations are similar for all furs, regardless of nature or origin, and these will be discussed briefly.

Inasmuch as the first great purpose of fur dressing is to render the skins more or less permanently immune from the processes of decay, it is necessary to prepare the pelts so as to be most fit to receive the preserving treatment. The skins as they are delivered to the fur dresser have, in the majority of cases, been stretched and dried to preserve them temporarily, while in some instances, especially with the larger furs like bears and seals, they are salted and kept moist. The flesh-side of the pelt still has considerable fleshy and fatty tissues adhering to it, and the hair is generally soiled and occasionally blood-stained. In order to get the pelts into such a condition that they can be worked and manipulated, they first have to be made soft and flexible. Very greasy skins are scraped raw in order to remove as much as possible of the attached fat, the operation being known as beaming or scraping. The typical beam, shown in Fig. 3, consists of a sloping table usually made of some hard wood, and placed at an angle of about 45°. It is generally flat, although in some instances convex beams are also used, about a yard long, 8 to 10 inches wide, and firmly supported at the upper end. The skin is placed on the beam, flesh-side up, and is scraped with a two-handled knife (Fig. 4), always in a downward direction.

Fig. 3. Beam.

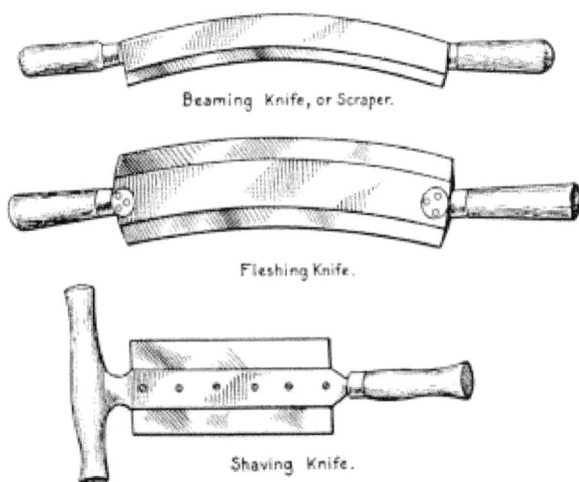

Beaming Knife, or Scraper.

Fleshing Knife.

Shaving Knife.

Fig. 4. Knives Used in Fur Dressing.

The first step in softening the skins is to get them thoroughly moistened, and this is variously done, depending on the nature of the skin. Lambs, for example, require the gentlest means of wetting them, while rabbits can stand soaking in water for several days. The manner and duration of moistening must be adjusted to the character of the pelt. For the putrefactive processes which were stopped by stretching and drying the skins, continue as soon as the pelt is again moistened. The progress of decay causes the evolution of certain gases, the simplest of which is ammonia, and eventually, if permitted to proceed, brings about the complete disintegration of the skin tissue. It has been found that a certain amount of gas formation is necessary to loosen up the fibres in order to get the best quality of leather after tanning. This process must be interrupted at the proper time and not allowed to proceed too far.

Skins which have been preserved fresh by salting, require only a comparatively short time (about 2 hours) to become softened by soaking in clean, soft water. Most dried skins need a longer treatment before they are sufficiently flexible. The addition of certain substances to the water facilitates and accelerates the softening. In some instances salt water is used for soaking the pelts, the preservative action of the salt tending to prevent any loosening of the hair. A solution of ¼% borax is very effective in rendering the skins soft, and clean as well. Borax has an exceedingly mild alkaline action, and causes a slight swelling of the skin tissue, which then absorbs the water more readily. Being also preservative and antiseptic, borax tends to prevent decomposition of the skin tissue. Another chemical of a different nature, but equally effective is formic acid, used in the proportion of 1.5–2.5 parts per 1000 parts of water. Formic acid also induces a swelling of the skin, the pelts being soaked in a short time, and the antiseptic action of the acid obviates the possibility of the hair becoming loose. The water used should be fresh and clean, and the soaking must be stopped as soon as the skins have become soft and flexible. Sometimes the skins are allowed to soak overnight in water, while in other cases, the pelts are just moistened by dipping in water until thoroughly wet, and then laying them in a pile for several hours, or overnight. Another method which is practised with certain types of skins is the use of wet sawdust or of sawdust moistened with salt water. The fur skins are either embedded in the sawdust or drummed with it for several hours, or until sufficient moisture has been absorbed to render them flexible. By this means there is no danger of the skins being over-soaked, or of the hair being loosened. When the skins have been properly wetted, they are drawn with the flesh-side across the edge of a dull knife-blade, in order to help loosen the texture of the skin. They are then put into a tramping machine and worked until completely softened. In the case of large or heavy skins, the moistened pelts are worked on the beam with a dull beaming knife to impart thorough softness and flexibility.

The pelts are then cleaned with particular reference to the hair. With some furs this is accomplished simply by drumming for several hours with dry sawdust, whereby the oil and dirt are removed from the hair, and the hair is then freed from the sawdust by caging. Other skins are washed, being passed through a weak soap solution for a short time, the dirty spots being brushed. Occasionally an extract of soap-bark is used in place of the soap, being even more effective. The cleansed skins are then thoroughly rinsed to remove any of the cleaning material, which would affect the gloss of the hair if allowed to remain on the skins. Then in order to eliminate as much as possible of the water in the skins, they are hydro-extracted, a centrifugal machine of the type shown in Fig. 5 being used. The basis of its action depends on the utilization of the principle of centrifugal force. The machine consists essentially of a perforated metal basket generally made of copper, capable of being rotated at a high speed. Surrounding the basket is an iron framework, polished or enamelled on the inside. The wet skins are placed in the rotating basket, fur side toward the perforations, and the water which is thrown off from the skins passes through the little holes, and is caught up on the walls of the outside frame, from where it is led off through suitable ducts. The centrifugal device is properly equipped with balancing and regulating attachments, as well as with a brake. The power may be applied by the over-drive or the under-drive as is most desirable in the particular case. The inner surface of the basket can also be enamelled or otherwise made resistant to the action of acids or other chemicals.

Fig. 5. Centrifugal Machine.

(*Fletcher Works, Inc., Philadelphia*)

When the skin is removed from the animal, as much as possible of the adhering fat and flesh is scraped off, but in spite of this, and in spite of subsequent beaming by the fur dresser, there is always a thin layer of flesh and fatty material remaining and this must be removed so as to expose the corium, enabling the efficient action of the chemicals used in the tanning processes. The process of removing this undesirable layer from the flesh-side is known as fleshing. It is a rather delicate operation, requiring considerable experience and dexterity on the part of the worker, for it is exceedingly easy to cut into the skin and damage the fur. A fleshing knife of the type commonly used is shown in Fig. 6. It consists of a sharp blade fastened at a slight angle from the vertical, with the cutting edge away from the workman, who straddles the bench, and by drawing the skins back and forth across the edge of the blade, removes all flesh and fat, leaving the corium free and clean. Large skins cannot conveniently be fleshed in this fashion. They are placed on the beam, and fleshed with a fleshing or skiving knife similar to the beaming knife, but consisting of a slightly curved, sharp two-edged blade having handles at both ends. Frequent attempts have been made to use suitable machines to do this work. A type of machine which has met with considerable success is depicted in Fig. 7. It is fashioned after the models used for the fleshing of hides for leather manufacture, and has special adjustments and regulating devices which afford protection for the hair part of the fur. From time to time other fleshing machines are put on the market, yet none of them seems to enjoy any great popularity, for fleshing is still largely a manual operation. With some classes of pelts, fleshing presents some difficulties, and chemical means have to be resorted to in order to loosen the flesh sufficiently to enable proper fleshing. In the case of large furs like bears, leopards, and the like, which while of no great importance in the fur trade, are occasionally met by the fur dresser, the skins after being soaked, and washed with soap-water, are partially dried; then the flesh-side is treated with technical butter or oil, which is tramped in. A mixture of salt water and bran is then applied to the skins, thereby causing a swelling action to set in, and the flesh becomes loosened, and is easily removed by fleshing on the beam. Seals receive a special treatment which makes them soft, and gives them greater stretch after they are tanned. A paste made by mixing a very dilute solution of caustic soda with an inert substance like French chalk, china clay, etc., is applied to the corium after the skins have been fleshed, then the pelts are folded up, and allowed to lie for several hours. They are then entered into a dilute solution of calcium chloride and left overnight. After being washed in a paddle or drum, first with fresh water, and then in water containing lactic or formic acid to remove the lime, the skins are ready for tanning.

Fig. 6. Fleshing Knife on Bench.

Fig. 7. Fleshing Machine.

(Turner Tanning Machinery Co., Peabody, Mass.)

CHAPTER V
FUR DRESSING
TANNING METHODS

After the pelts have gone through the preliminary operations of softening, washing and fleshing, they are ready to receive the treatment which will convert the easily decomposing skin into leather of more or less permanency, depending on the method used.

During the past century, considerable study has been made both by scientific and technical people, of the problem of leather formation. Numerous theories as to the nature of the process have been evolved, but even to this day, no satisfactory explanation has been given which would account for all the facts as they are now known, so the matter is still a subject of considerable controversy. Procter, who is one of the leading authorities on leather today discusses the development of the tanning theories as follows:

"The cause of the horny nature of dried skin is that the gelatinous and swollen fibres of which it is composed not merely stiffen on drying but adhere to a homogeneous mass, as is evidenced by its translucence. If in some way we can prevent the adhesion of the fibres while drying we shall have made a step in the desired direction, and this will be the more effective the more perfectly we have split the fibre-bundles into their constituent fine fibrils, and removed the substance which cements them. The separation of the fibres can be partially attained by purely mechanical means.... Knapp, to whom we owe our first intelligible theories of the tanning process, showed that by physical means the separation and drying of the fibres could be so far effected as to produce without any tanning agent a substance with all the outward characteristics of leather, although on soaking it returned completely to the raw hide state. He soaked the prepared pelt in absolute alcohol, which penetrated between, and separated the fibres and at the same time dried them by its strong affinity for water. More recently, Meunier has obtained a similar result by the use of a concentrated solution of potassium carbonate which is even more strongly dehydrating.

"Knapp made a further step by adding to his alcohol a small quantity of stearic acid which, as the alcohol evaporated, left a thin fatty covering on the fibres which completely prevented their adhesion, and reduced their tendency to absorb water; and he so produced a very soft and white leather. Somewhat similar are the principles of the many primitive methods which apply fatty and albuminous matters, grease, butter, milk, or brains to the wet skin, and by mechanical kneading and stretching, aided by capillarity, work these matters in between the fibres as the water evaporates. Such methods are still used, and enter into many processes in which other tanning agents are also employed.

"Building upon these facts, Prof. Knapp advanced the theory that the effect of all tanning processes was not to cause a change in the fibres themselves, chemical or otherwise, but merely to isolate and coat them with water-resisting materials which prevented their subsequent swelling and adhesion. True as this theory undoubtedly is in many cases, it can hardly be accepted as the whole truth, and it seems incontestable that frequently the fibres themselves undergo actual chemical changes which render them insoluble and nonadhesive.

"Before Knapp's work, the prevalent theory, at least as regards vegetable tannage, had been a chemical one, started by Sir Humphrey Davy. If a solution of gelatine be mixed in proper proportion with one of tannin, both unite to form a voluminous curdy precipitate; and, according to Davy's ideas, this was amorphous leather. Against this, it was urged that even the supposed 'tannate of gelatine' itself could not be a true chemical compound, since the proportions of its constituents were considerably varied by changes in the strength of the solutions, or by washing the precipitate with hot water; and further, that in chemical compounds, the form was changed, and no trace of the original constituents appeared in the compound; while in leather apart from some change of color and properties, the original fibrous structure remained unaltered.

"This reasoning appears much less conclusive now than it did in Knapp's day. Against the last objection guncotton may be quoted as an instance of profound chemical change with no alteration in outside appearance; and it is recognized that, especially among complex organic substances, chemical reactions are rarely complete, but that stable positions are reached, so-called 'equilibria,' in which the proportion of changed and unchanged substance is dependent on concentration or other conditions; and that therefore such a precipitate might well be a mixture of gelatine with a true gelatine tannate from which further portions of tannin might be dissociated by water.

"With the clearing up of old difficulties, however, the conflict between chemical and physical theories has, as is usually the case, merely passed into a new phase. Years ago, it was shown by Linder and Picton and others, that liquids could be obtained which were not really solutions of ions or molecules, but merely suspensions like that of clay in water, or butter-fat in milk; but so finely divided as to appear clear and transparent, and pass through filters like true solutions. Later, by means of the ultra-microscope their discrete particles have actually been made visible, each of them consisting of many molecules of the suspended substance. Nevertheless, these particles have many molecular properties, possessing plus or minus electrical charges; behaving like large ions under the influence of an electrical current; and mutually precipitating and neutralizing each other when positive and negative are brought together. Such solutions are called 'colloid,' and those of gelatine and tannin are of the class, so that it is now often said that the precipitation of gelatine by tannin, and the fixation of tannin by gelatinous fibre are merely 'colloidal' and 'physical,' and not 'chemical' phenomena. Admitting the facts, the question still arises whether the distinction between chemical and physical is not here one without a difference; and whether between the purely ionized dilute solution of a salt and the coarsely granular clay suspension there is any point where a definite line of demarcation can be drawn. The writer inclines to the view that there is not; and that ionic and colloidal combinations are extreme cases of the same laws, both physical, and both chemical."

There are several methods which are used in tanning furs, each having its peculiar characteristics and qualities, and possessing individual advantages and disadvantages. In order to be able to judge the merits of the various processes, it is necessary to have a criterion which can serve as a basis of reference. Fahrion, a recognized authority and investigator in this field, gives a definition of leather which is usually accepted as a standard for comparison. He says: "Leather is animal skin, which on soaking in water and subsequent drying does not become hard and tinny, but remains soft and flexible; which does not decay in the presence of cold water; and which does not yield any gelatine on boiling with water." While the requirements set forth in this statement are essential for leather, and a compliance with them would also be desirable for tanned furs, a somewhat less rigorous standard of conditions to fulfil is satisfactory for the general needs and purposes of furs. The chief qualities which tanned furs must possess, with particular reference to the leather side of the pelt, are retention of softness and flexibility after being moistened by the furrier for manufacturing purposes, and subsequent drying; and freedom from a tendency to decay during this operation and thereafter. If the furs are to be dyed, the effect of the dyeing must also be considered, and the tanning must be such as to enable the dyed furs to possess the above qualities.

The most important tanning processes employed for furs are the following:

- 1. Salt-acid tan, or pickle.

- 2. Mineral tans.

- 3. Chamois tan.

- 4. Formaldehyde and similar tans.

- 5. Combination tans.
- 6. Vegetable tan.

1. SALT-ACID TAN, OR PICKLE

This is one of the most extensively used methods for tanning furs, and is also very cheap and easily applied. A typical formula for this tan is the following: A solution of salt is prepared containing about 10% of common salt, sodium chloride, and to this is added $\frac{1}{2}$–$\frac{3}{4}$ ounce of sulphuric acid for each gallon of tanning liquor. The proportions may be varied within certain limits, but the figures here given are those which have proven successful in practise. The solution should be made in a wooden or earthenware container, free from any metal, as it would be attacked by the acid. The liquor is then applied to the flesh-side of the fleshed skins by means of a brush, making sure to touch all parts of the pelt. They are then placed in a pile and allowed to remain thus until tanned, an operation which occupies a time ranging from a few hours to two or three days depending on the thickness of the skins. When the corium has lost its translucence and has become of a milky-white color throughout the entire thickness of the skin, as can be seen by viewing a cross-section, the skin may be considered tanned. In some instances, where the hair of the fur can stand immersion without injury, the skins are entered into the pickling solution and allowed to remain for 12 to 24 hours, which is generally a sufficient time to tan them in this manner.

The acid of the pickle causes the skin to swell, the salt then penetrating between the fibres of the corium, and at the same time reducing the swelling of the skin. The acid also neutralizes the alkaline products of decomposition which may form, while the salt acts as a deterrent to the progress of the putrefactive processes. When the skin is dried after tanning, and stretched and finished, a soft white leather is obtained which is permanent as long as it is kept dry. It is the salt which causes the fibres of the skin to be completely differentiated and thus prevents their adhesion.

It is interesting to note that other acids besides sulphuric can be used for the pickle, organic as well as mineral, formic acid in $\frac{1}{4}$% solution being especially effective and giving excellent results, but is more expensive than the mineral acid. A method, which in principle is identical with the pickle, but carried out in an entirely different manner, is the lactic acid fermentation process, or "Schrot-beize" as it is called in German. The procedure is in general as follows: "The fleshed skins are placed on tables, flesh-side up, and covered with a layer of bruised barley grains, or a mixture of 3 parts of wheat bran and 2 parts of rye flour. Then the head, tail and legs are turned inward, and the skins rolled up in little cushions, hair-side out, and placed in a vat. When this is filled with the skins a solution of common salt is poured over them, and they are allowed to remain thus in a moderately cool place for 24 hours. After this time, the skins are carefully unrolled, so as not to remove any of the adhering solid materials, and turning the skins hair-side inward, they are laid flat together in pairs and placed in an empty vat. After another 24 hours they are again unpacked and replaced in another vat, care being taken each time to keep all the solid particles adhering to the flesh-side. This operation is continued and repeated until the skins are properly tanned, which takes from 10 to 14 days, depending on the weather and the temperature. The skins are then removed, rinsed free of the tanning substances, pressed, dried and finished." A somewhat modified form of this process is the so-called Russian tan, which is usually done in the following manner: 5 parts of bruised barley grains are mixed with ten parts of luke-warm water in a vat, which is then covered up. A small quantity of brewers' yeast is also added to aid in the fermentation. As soon as the mixture develops a slight heat, one part of fresh whey is added, and the fleshed skins entered into the tanning liquor in which they remain for about 12 hours. They are then tramped in the mixture so as to effect greater penetration, and left until the tanning process is complete. Whey is the milk fluid left after the casein and most of the fat have been removed from the milk by coagulation, and consists practically of a solution of all the milk-sugar or lactose, and the lactic acid of the milk, together with a small percentage of mineral salts, and a slight amount of fat. By fermentation, the milk-sugar is converted into lactic acid, which helps to effect the tan by swelling the skin.

The effectiveness of the fermentation processes depends to a considerable degree on the action of certain bacteria and yeasts. Bacteria are one-celled organisms belonging to the vegetable kingdom, and some are so small as to be scarcely visible under a microscope, while some indeed cannot be seen by any means, their existence being inferred from their effects. As they vary in size, bacteria also vary in shape, some being spherical, others in the form of long, thin rods, while still others are of a spiral shape; another common form is the dumb-bell shaped bacterium. Some types are provided with what are known as flagella, which resemble fine hairs attached to the body of the organism, and which enable it to move about actively in liquids. The food of bacteria is always in liquid form, as only in this condition can it be absorbed. However, some kinds of bacteria attack solid substances from which they obtain their nourishment, but this is done in an indirect way, by secreting certain fluids known as enzymes, which dissolve or digest the material and convert it into a form that can easily be absorbed by the bacteria. The enzymes are non-living chemical substances, which possess the peculiar property of bringing about the chemical change of an almost indefinite amount of material upon which they act, without themselves being in any way changed. Yeasts also act in a manner similar to the bacteria in causing various chemical changes, particularly inducing fementations. In the simple "Schrot-beize," the starch contained in the bran or barley grains is first converted to a soluble sugar by means of enzymes secreted by the bacteria which are always present. This sugar then undergoes an acid fermentation, with the formation of lactic and acetic acids, due in this case to organisms known as the *bacterium furfuris A* and B. The action of the Russian tan is similar, but quicker. In this case, the sugar is already present in soluble form, and the yeast cells cause its fermentation with the production of lactic acid. In both cases, the acids as they form swell and loosen up the skin fibres slowly, the salt penetrating between them, and keeping them separated on drying. Both methods give results which are equally good, but by the Russian tan the skins acquire a disagreeable odor, which makes this method of dressing objectionable.

The lactic acid fermentation processes have an advantage over the pickle, in that the slow formation of weak organic acids with their gradual action produce a softer leather, with a gentler 'feel,' the presence of the flour and the grains of the tan, aside from their tanning action, contributing to the fullness and softness of the leather. There is also less likelihood of the leather being subsequently affected by the presence of the acid in it, as lactic and acetic acids are much less injurious than sulphuric acid to leather. These disadvantages of the pickle can to a large degree, be overcome without any great difficulty. On the other hand, the matter of the length of time of the tanning process, shows the acid pickle at a great advantage, and so, especially for furs other than those obtained from sheep and goats, the pickle is in most cases used as the principle method of tanning. In Austria, Russia, and to a certain extent in Germany also, the "Schrot-beize" is still considerably employed, chiefly for dressing sheep and lamb skins. The dressing of the various kinds of Persian lambs, caraculs, astrachans, etc., in the native center of the industry in Buchara and surrounding districts, is also a "Schrot-beize," barley, rice flour or rye flour, and salt water being used to prepare the skins, the manipulations being essentially the same as those described above, although carried out in cruder and more primitive fashion.

2. MINERAL TANS

The basis of the tanning of furs by means of solutions of mineral compounds is the fact that the basic salts of certain metals are capable of producing leather. It has been found that compounds of aluminum such as alum or aluminum sulphate, or any other soluble neutral salt of aluminum, possess tanning powers. Other metals which are capable of forming salts of the same type are also endowed with the quality of converting skin to leather under suitable conditions, chromium and iron being the most important metals in this connection. Chemically these metals all belong to the same group, and have properties which are very similar in many respects, the characteristic of most importance for tanning purposes being the quality of forming soluble basic salts by the addition of alkalies or alkaline carbonates to solutions of their neutral salts, or in certain instances simply by the action of water upon these neutral salts. By neutral salts are meant those in which the metallic content is combined with the normal proportion of acid; while basic salts are those in which the acidic portion is present in less than the normal ratio, being partially replaced by a hydroxide group. When the acid part of the salt has been entirely replaced in this way, the compound is called a hydroxide or hydrate of the metal. Between the neutral salt and the hydroxide several different basic salts are possible, some being soluble, while others are insoluble. If into a solution containing a basic salt of either aluminum, iron or chromium a skin be entered, a part of the basic salt will be precipitated on it in insoluble form. Inasmuch as neutral salts of these metals when dissolved in water split up to a small degree, into free acid and soluble basic salt, a skin immersed in such a solution will also absorb the basic salt in an insoluble form. Upon these facts in general, depends the action of the mineral tans used in tanning furs.

A. *Alum Tan*

The alum tan is one of the oldest methods of producing leather, being employed by the Romans about two thousand years ago, and it is believed, by the Egyptians at a much earlier period. Its extensive use in Europe, however, dates from the time of the conquest of Spain by the Moors, who introduced the process.

At the present time, rabbits and moles are tanned by this process, as are also at times other furs such as muskrats, squirrels, sables, martens, etc., when a better tan is desired than that produced by the pickle. Ordinary alum, which is a double sulphate of aluminum and potassium, and aluminum sulphate are the chief compounds used for this tan. In recent years, the aluminum sulphate has to a considerable degree replaced the alum for tanning, inasmuch as it can be cheaply obtained in a sufficiently pure form, and contains about one and one-half times as much active aluminum compound as does alum.

While the aluminum salt can be used alone for tanning, it produces a stiff, imperfect leather, so salt is always added. The ratio of the salt to the aluminum sulphate or alum can vary within rather wide limits, the mixtures used in practise ranging from one part of salt to four parts of the aluminum compound, up to equal parts of both, or even in some formulas, a greater proportion of salt than of the other constituent. Ratios which are most common are four of alum to three of salt, or two of alum to one of salt.

When aluminum sulphate is dissolved in water, a small part of it splits up into a soluble basic salt and an equivalent amount of free acid. The reaction may be shown as follows:

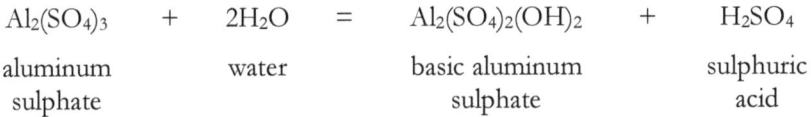

$$Al_2(SO_4)_3 \quad + \quad 2H_2O \quad = \quad Al_2(SO_4)_2(OH)_2 \quad + \quad H_2SO_4$$

| aluminum sulphate | water | basic aluminum sulphate | sulphuric acid |

When a skin is entered into such a solution, the free acid is absorbed, causing a swelling of the pelt. While this is taking place, a further quantity of the neutral aluminum salt splits up into more basic salt and free acid. At the same time the basic aluminum sulphate is also taken up by the skin, probably attaching itself to some of the acidic groups contained in the skin substance, in a manner analogous to the combination of the acid with the basic groups of the skin substance. A point is reached, however, when the skin is no longer able to take up more of the basic salt, for the presence of the acid undoubtedly acts as a deterrent. The skin, if dried after such a treatment contains a small amount of aluminum, which is insufficient to tan the pelt properly, and as a result this comes out in an undesirable and quite useless condition. If to the solution of the aluminum sulphate salt is added, a different result is obtained. To a certain extent the salt acts here as in the pickle. The skin on absorbing the free acid of the solution naturally swells, but the salt reduces this swelling, and at the same time, by penetrating between the fibres and dehydrating them, produces a leather as in the pickle. In addition, the presence of the salt enables a greater amount of basic aluminum sulphate to be formed, and thus a greater quantity is taken up by the skin. On drying and stretching after such a treatment, a soft, flexible and stretchable leather is obtained.

The number of formulas for tanning furs by this process is legion, the principle being the same in every instance, and mixtures of salt and alum or aluminum sulphate form the basis of the various tans. Following are a few typical formulas, which have been found to be of practical value:

A solution is prepared by dissolving 7.5 lbs. of alum and 3 lbs. of common salt in 20 gallons of water. When cool, the clean, fleshed skins are entered, being paddled or drummed for a short time and then allowed to remain until tanned. By this method the hair also takes up some of the alum, and if the skins are to be dyed, unevenness may result. In order to avoid this, the tanning may be effected by brushing a stronger solution on the pelt. A mixture of 4 lbs. of alum and 3 lbs. of salt, dissolved in 8 gallons of water, and made into a paste by the addition of 4 lbs. of flour, is applied to the flesh-side of the skins. These are then placed in pairs, flesh-side together, and allowed to remain in a pile until tanned. Sometimes a second application is given. The flour may be omitted, but it serves to cause the tanning mixture to adhere better to the skins.

Still another method is the following: Into the flesh of the moist, fleshed skins is rubbed a mixture of two parts of dry powdered alum with one part of salt. After allowing time for it to be absorbed, another application is given, rubbing in well, and especially treating the thick parts. The pelts are then folded up, or rolled together, flesh-side in, and placed in a vat or tub, which is covered up to prevent drying. They are left so until tanned, as shown by examination and test. They are then rinsed, hydro-extracted and dried, and after stretching and finishing, a soft, white, pliable leather is obtained.

B. *Chrome Tan*

By using chrome alum instead of ordinary alum, together with salt, skins can be tanned, but the leather formed is not altogether satisfactory. The basic principle here is the same as in the alum tan, depending on the formation of soluble basic chrome sulphates in the solution of a neutral sulphate. The method employed at the present time, the so-called one-bath process as distinct from the two-bath process, which cannot be applied for tanning furs, involves the production of the basic chrome sulphate by the addition of an alkali or an alkaline carbonate to the solution of the neutral salt. It was Prof. Knapp who first published this process as early as 1858; but it was not until 1893 that it was shown to be of practical value, and was then patented in this country by Martin Dennis. Since that time it has been in general use with but slight modifications.

The chrome tan is used only to a limited extent in the tanning of furs, the method requiring very careful treatment and accurate supervision during the various stages of the process, and the leather coming out colored a pale-blue-green tint, which for some purposes is objectionable. In some plants ponies and rabbits are tanned with chrome; and when the skins are to be dyed by means of certain coal tar dyes, they have to receive a chrome tannage. The leather produced by a chrome tan is very durable, and possesses great resistance to the action of water.

Any salt of chromium, with either mineral or organic acids, can be used, but chrome alum is the one most commonly employed. If a skin is entered directly into a solution of a chrome salt made basic with an alkali, the precipitation of the insoluble basic salt will take place very rapidly, and the tanning will be only superficial. The procedure is therefore first to treat the skins with a chrome solution which forms only small quantities of the basic salt. After the skins are impregnated with the solution, this is made basic, so that the real tan will take place within the skin tissues among the fibres of the corium. A common formula is the following: 5 lbs. of chrome alum are dissolved in 10 gallons of water. The skins are entered into the solution at about 70° F. and paddled for about 2 hours, or drummed for one hour. Then a solution of three pounds of washing soda is added slowly to the liquor which is then stirred up well, and the skins drummed or paddled again for an hour or two, and then left in the liquor for 12 to 24 hours till completely tanned. The skins are rinsed, and washed in $\frac{1}{2}\%$ solution containing $\frac{2}{3}\%$ of the weight of the skins of borax. The pelts are then well washed in clean water, hydro-extracted and dried.

C. *Iron Tan*

Tanning by means of iron salts has thus far been merely a matter of scientific interest and has not found any practical use. The principle involved is identical with that of the preceding mineral processes.

3. CHAMOIS TAN

The chamois dressing, as previously noted, is undoubtedly the oldest method of preparing leather from skin, the various fat-containing substances derived from animals, fish, birds, etc., being used for the purpose. The chief object of the fat was to coat the fibres of the skin, thus preventing their adhesion, and at the same time rendering them resistant to water. In the true chamois tan, the fat seems to have also a chemical function in contradistinction to the other which is merely physical or mechanical. For, if skins tanned by the chamois process be treated with a weak solution of an alkali, all the fatty materials should be removed thereby, but this happens only to a small extent, the pelt retaining its softness and pliability, and the other characteristic qualities of leather, indicating that the fat is combined intimately with the skin substance in a permanent fashion.

In tanning furs, various oils and fats are used, but not all are capable of producing a chamois tan. Among the fatty materials are mineral oils, and vegetable and animal oils and fats. Mineral oils are the distillation products of petroleum, partially liquid, and partially solid. Being inert substances, they have no tanning effect, but serve merely as water-proofing or fattening materials. Except for their oily nature they have nothing in common with fats, being quite unaffected by solutions of alkalies or of acids.

Vegetable and animal fats and oils are, when pure, neutral substances formed by the combination of fatty acids with glycerine. They possess the property of saponification, that is, of forming a soap when treated with an alkali, the soap being the alkaline salt of the fatty acid. Under certain conditions, the fat can be split up into free fatty acid and glycerine by the action of acids, or even water alone. Some fats on long standing, split up in this way spontaneously in the presence of moist air. As a general rule, those fats which exhibit this property to a marked degree are affected by contact with the air, due to the absorption of oxygen which reacts chemically with the fats, forming what are known as oxy-fatty acids, usually less soluble, and having a higher melting point than the original fats. Vegetable and animal fatty materials are classified on the basis of this phenomenon of absorbing oxygen from the air, those possessing this quality to a great degree being called "drying oils," others being "partially drying," or "non-drying." Olive oil, castor oil, cocoanut oil and cottonseed oil are examples of non-drying or partially-drying vegetable oils, linseed oil being the most important drying-oil in this class. Tallow, lard, butter-fat, neats-foot oil are non-drying animal fats, the drying oils being seal oil, whale oil, and cod-liver oil.

Fig. 8. Tramping Machine or "Kicker."

(*F. Blattner, Brooklyn, New York.*)

For tanning purposes, this property of absorbing oxygen is important, because only with drying oils can a true chamois tan be obtained, non-drying oils acting like mineral oils only as water-proofing materials. The details of the chamois process are not quite clear, there being considerable difference of opinion on the matter. But all the studies on the subject tend to prove that there are at least two phases to the process: first, the mechanical covering of the fibres with the fat, this property being common to all fats or oils which may be used; and second, the combination of the fat with the skin in some chemical way, as a result of the oxidation of the fat, a characteristic found only in the drying oils. During the oxidation of the fats, the glycerine in them is converted to acrolein or acryl-aldehyde, which also aids the tanning. It was at one time supposed that the tanning action was due to this aldehyde alone, but a chamois tan can be made with fatty substances from which all the glycerine has been removed. The evidence on this question, however, is not quite conclusive.

In general, the procedure of the chamois tan is as follows: The hydro-extracted, fleshed skins are rubbed on the flesh-side with a good quality of seal-oil. They are then folded up, and put into a 'kicker,' where they are tramped for two or three hours to work in the oil. The kicker is a machine such as shown in Fig. 8 consisting of a receptacle for the skins, and two wooden hammers which work up and down mechanically, turning and pounding the skins. (As many as 1000–1500 skins of the size of musk-rats can be worked at one time in such a machine.) The pelts are then taken out and hung up in a warm room for several hours, considerable oxidation taking place. Another coat of oil is then applied, which is again tramped in, and the skins are hung up once more and exposed to the air to cause the oil to oxidize. After the skins are sufficiently tanned they are rinsed in a weak soda solution to remove the excess oil, washed and dried. When skins with fine hair such as marten, sable, mink, etc., receive a chamois tan, they are not tramped in kickers as the delicate top-hair will be broken, and the value of the skin thereby reduced. Instead they are placed in small drums, together with metal balls of varying sizes and weights depending on the particular fur treated, and the oil is worked in by rotating the drum. Such a ball-drum, as it is called, is shown in Fig. 9.

Fig. 9. Ball Drum.

(*F. Blattner, Brooklyn, New York.*)

In conjunction with the chamois tan may be discussed the process of oiling, inasmuch as the method of application and the effect are both similar to the chamois tan up to a certain point. It is customary to treat skins tanned by any other method but the chamois process, with some oil in order to render them more impervious to water. The greatest variety of oils and fats can be used, the action in most cases being simply the mechanical isolation of the skin fibres by such a substance, thus corresponding to the first or physical phase of the chamois tan. The chemical phase, if it takes place at all, is usually slight, and is merely incidental. Oiling is generally applied either before drying after tanning, or after drying, the oiled skins being placed in a kicker and tramped to cause the oil to penetrate. In some instances the oiling material is put in the same mixture as the tanning chemicals, and the tanning and oiling are effected simultaneously.

Among fatty substances used for oiling are mineral oils, such as paraffine oil, and vaseline; animal fats, like train oils, butter, egg-yolk, glycerine, neatsfoot oil; vegetable oils, like olive oil, castor oil, cottonseed oil; also sulphonated castor oil and sulphonated neats-foot oil. These may be used singly or in various mixtures, an emulsion of an oil and a soft soap also being frequently employed.

4. FORMALDEHYDE TANS

Formaldehyde has proven to be of great value in the tanning of furs, usually in conjunction with other processes. Formaldehyde is a gas with a strong, irritating odor, and its 40% solution, which is the customary commercial form, also possesses this quality. When skins are treated for several hours with a very dilute solution of the commercial product, a leather is obtained which combines the properties of the alum tan and the chamois tan. Moreover, in the majority of observed cases, where furs have been tanned with formaldehyde, the skins seem to acquire a certain immunity to the attacks of vermin and moths. Although the skins do not in any way retain the odor of the formaldehyde, nevertheless these destructive agents seem to be repelled.

Numerous processes have been devised which use formaldehyde in connection with other substances for tanning. Thus in a German patent is described a method involving the alternate or simultaneous treatment of pelts with solutions of formaldehyde and alpha or beta naphthol. Both the formaldehyde and the naphthol exercise tanning actions, but the process is not used in practise.

In 1911, Stiasny, a well-known leather chemist produced a synthetic substance by the condensation of formaldehyde with a sulphonated phenol, forming an artificial tannin. This chemical, called "Neradol D," exhibits many of the properties characteristic of true tannins, although in no way related by structure and composition. By the use of "Neradol D" a soft, white and flexible leather is obtained, and it is therefore a suitable tanning material for furs.

5. COMBINATION TANS

In many instances more than one method is employed in tanning the furs, and in this way what is known as a combination tan is produced. While the various individual processes described give more or less satisfactory results by themselves, they generally possess some features, which for certain purposes may be undesirable, and which can be eliminated or considerably reduced by using other processes at the same time or subsequently. Some of the combination methods are, pickle with chrome tan, alum tan with chrome tan, and formaldehyde tan with pickle, mineral tan or chamois tan. By means of such combinations various qualities of tanned furs can be obtained, and if it is desired to produce a pelt having certain special characteristics, this can be brought about by combining two or more standard methods.

Some illustrations of combined tannages are the following: Alum-chrome tan. The skins are tanned by the regular alum process, then the constituents of the chrome tan are dissolved directly in the same bath, and the chrome tan is effected as usual. Chrome-formaldehyde tan. To the regular chrome tan solution is added ½ lb. of formaldehyde for every 10 gallons of chrome liquor. The rest of the process is as ordinarily.

6. VEGETABLE TANS

In practise, the vegetable tanning matters are not used for furs, although in some special instances gambier cutch may be employed occasionally with some other tan. However, many of these tannins also have dyeing properties, and are used in dyeing the furs. In this connection it must be mentioned that furs dyed with these materials also receive a vegetable tan, which improves the quality of the leather to a considerable extent.

Comparison of The Various Tanning Methods

In choosing a method for tanning any particular kind of fur, several factors must be considered. The nature of the pelt, insofar as it is weak or strong; the time, labor and cost of materials required by the tanning process; the effect on the leather of the different dyes and chemicals used in dyeing, if the skin is to be dyed, are a few of the points requiring attention and consideration.

For furs which are only to be dressed, a simple tan like the pickle will suffice in most cases. Special instances, such as the rabbit and mole already mentioned, and a few other furs are tanned by the alum method. The pickle is undoubtedly the cheapest and simplest method of tanning skins, and yields a soft, white leather which is permanent as long as it is kept dry. If it is put into water, about 25% of the salt contained within the pelt dissolves out, and the acid present swells up the tissues. If the skin is dried in this condition, it will come out hard and brittle, tending to crack very easily. By treating the leather before drying with a strong salt solution, a good deal of the extracted salt will be replaced, and on drying and stretching, it will work out soft. Skins tanned by the "Schrot-beize" are affected by water in quite the same manner as the pickled skins.

The alum tan gives a leather similar to that produced by the pickle, but with the advantage that the skins possess greater stretch and flexibility. In its resistance to water, the alum-tanned pelt is quite as susceptible as the other. As a general rule, the skin absorbs about 6% of its weight of alum from the tanning solution, but gives up three-quarters of this when it is soaked in water, producing on drying, a hard, stiff leather. The chrome tan is especially impervious to water, easily resisting temperatures of 80° C., and even boiling water. It is employed to only a limited extent on account of the special effort and care required to obtain satisfactory results, also because the pelt acquires a pale blue-green color which is not desired on dressed skins. The chamois tan, and some of the combinations of the formaldehyde tan with the other methods, give very soft, flexible leathers which possess a sufficiently great resistance to the effects of water and heat.

In tests made to determine the best working temperatures for dyeing skins dressed by the salt-acid tan, and for skins dressed by the chamois process, some very interesting facts were brought out. These two tans were chosen because they represent opposite extremes, the salt-acid tan usually giving the poorest results, and the chamois tan giving the best results in practise in dyeing. Other methods, except the chrome, range between these two. The procedure in these experiments was to treat the skins at ordinary temperatures in water, or dilute solutions of the various chemicals and dyes usually employed in dyeing, and then heat these solutions until the leather just began to shrink and shrivel up. This point, called the shrinking point (S.P.), gave the temperature to which the skins could be subjected in the given solution without danger to the pelt. (The experiments and observations were made by Erich Schlottauer, while director of a large German fur dressing and dyeing plant).

The first observation made was that different furs tanned by the same process were affected differently in the same solutions. Thus in plain water, three furs, all tanned by the acid-salt tan, had shrinking points varying by several degrees; similarly with two different furs tanned by the chamois process, there was a variation in the shrinking point of two degrees. The explanation of this discrepancy among the different skins may be that there was a slight difference in the conditions under which they were tanned, experiments showing that a maximum difference of 4° C. may exist among skins tanned by the same process, but not under the same or identical circumstances. Another reason for the variation may be the fact, that some skins are more greasy than others, and are thus more resistant to the effects of water or of some chemicals. The furs with the higher shrinking points in water were those which naturally are more greasy than the others.

Weak solutions of acids tend slightly to lower the shrinking point, while weak solutions of alkalies appreciably raise it, in both chamois-tanned and salt-acid-tanned skins. Solutions of dyes and mordants as a general rule increase the resistance of the skin to heat, varying quantities of these substances having no, or little different effects on the shrinking points. Previous treatment of the leather with some oil considerably raises the shrinking point of the pelt. Formaldehyde effects a great increase of the resistance of the skins to heat, especially with chamois-tanned furs. The experiments in this case were made by first treating the skins in the weak formaldehyde solution, and then determining the shrinking point in plain water.

Two skins, both dressed by the "Schrot-beize," a Persian lamb and an astrachan, after dyeing had shrinking points almost 10 degrees higher than when undyed. The extra tannage which the skins received from the tannins used in the dye mixtures for these furs, accounts for this increased resistance to heat.

The following tables give the observed figures in the different experiments:

TABLE I

	A S.P.	B S.P.	C S.P.
Salt-acid Tan	C.	C.	C.
Australian Opossum	46°	58°	45°
Marmot	45°	50°	42°
Skunk	47°	56°	43°
Chamois Tan			
Mink	52°	61°	45°
Muskrat	50°	58°	42°

A—Water

B—Water plus 1% Ammonia (s.g. 0.910)

C—Water plus 1% Sulphuric acid (66° Beaumé)

TABLE II

	A S.P.	B S.P.	C S.P.
Salt-acid Tan	C.	C.	C.
Australian Opossum	53°	52°	54°

Chamois Tan			
Mink	59°	59°	59°

A—1000 c.c. water plus 40 c.c. Peroxide plus 5 c.c. ammonia

B—500 c.c. water plus 2 grams Ursol D (Para-phenylene-diamine)

C—500 c.c. water plus 5 grams Ursol D

TABLE III

	A S.P.	B S.P.	C S.P.	D S.P.
Salt-acid Tan	C.	C.	C.	C.
Australian Opossum	51°	51°	53°	56°
Chamois				
Mink	59°	59°	61°	62°

A—500 c.c. water plus 5 grams ground nut-galls

B—300 c.c. water plus 2 grams pyrogallic acid

C—500 c.c. water plus 2 grams potassium bichromate

D—Water, after treating leather with rapeseed oil

TABLE IV

	A S.P.	B S.P.	C S.P.	D S.P.
Salt-acid Tan	C.	C.	C.	C.
Australian Opossum	49°	49°	55°	50°
Chamois Tan				
Mink	59°	67°	69°	70°

A—500 c.c. water plus 5 c.c. formaldehyde for 1 hour

B—500 c.c. water plus 5 c.c. formaldehyde for 12 hours

C—500 c.c. water plus 10 c.c. formaldehyde for 3 hours

D—As in C, but treated with 500 c.c. water plus 5 c.c. ammonia, instead of water alone.

TABLE V

	A S.P.	B S.P.
	C.	C.
Persian Lamb	44°	54°
Astrachan	47°	55°

A—Before dyeing

B—After dyeing

As a result of these experiments it may be concluded that the maximum temperature for drying salt-acid tanned skins should be 40° C., while for chamois tanned skins the temperature may be permitted to reach 45° C. without any danger of the leather being affected. Moreover, in the case of pickled skins, the matter of extraction of the tanning agent, as well as that of the leather becoming "burned" may be effectively counteracted by brushing some oil or fat on to the leather side before dyeing the pelt.

The shrinking points of skins dressed by the various tanning methods are constant within certain limits, depending on the nature of the skin and on the conditions of tanning, and it is possible by observing the shrinking point, in conjunction with other characteristics of a given pelt, to determine what method of tanning was used.

CHAPTER VI
FUR DRESSING
DRYING AND FINISHING

One of the most important operations of all the fur dressing processes is the drying of the skins. For even when all the previous steps have been successfully completed, there is still a great possibility of the skin being injured if the drying is not properly and carefully carried out.

The essential requirements for good drying are proper temperature, uniformity and rapidity. The leather part of the fur cannot, in the moist state, resist temperatures exceeding about 45° centigrade, for when dried, the skin turns out hard and stiff, and cracks easily. The furs must therefore be dried at an initial temperature of 25° to 30° centigrade, and as the moisture is gradually removed, the temperature may be raised, for the less water that remains in the pelt, the less is the leather affected by the heat, and the more difficult is the removal of its aqueous content.

If the drying process is not a uniform one, that is, if all the skins in a lot are not subjected to the same drying conditions, then after the drying has proceeded for a certain time, some skins may be quite dry while others are not, or there may be as many different degrees of dryness as there are skins drying. There is also the possibility of great variation in the amount of moisture removed from different parts of the same skin. Such a state of affairs requires an extra expenditure of time, labor and heat power in order to get the whole lot of furs into a more or less uniform condition. Moreover in some kinds of furs, especially those with thick skins, when the drying is not even, there is danger of the epidermal layer drying away from the corium, and subsequently peeling and cracking. Uniformity of drying requires the maintenance of a reasonably constant temperature equally distributed throughout all parts of the space where the drying is done, so that all the furs may be dried under the same conditions.

Rapidity of drying is desirable not only because it is beneficial to the condition of the pelt, but also from the point of view of practical business economy. The space occupied by the drying should be as small as possible compatible with the volume of work, and with the efficiency of operation. Slow drying involves the use of much space to take care of all the skins to be dried, or an accumulation of pelts ready to be dried, neither of which conditions is efficient or desirable.

It was formerly the general custom, still practised in some establishments, to dry the skins by hanging them up, leather-side out on lines in a large room or loft, the heat being usually supplied by steam pipes. Such a procedure occupied often as long as two or three days to get complete drying, involved a great deal of labor, and the results were far from uniform. In fact, in order to get the skins more nearly equable, it was necessary to subject them to an additional operation. This usually consisted of rotating the skins in a closed drum for several hours, the constant intermingling of the pelts in contact with each other causing any moisture left in them to be evenly distributed throughout the whole lot. The skins, by this process also are rendered somewhat softer and more flexible, but by drying under proper conditions the entire extra operation can be dispensed with, the furs coming out quite as soft and flexible without the drumming.

A great improvement was the adoption of large fans to circulate the heated air in the loft, thereby approaching more nearly an even temperature. More modern devices have, however, been developed, whereby drying can be effected in the most uniform manner, with perfect control of temperature, and requiring the least possible consumption of space, time, labor and power. A typical arrangement consists of a large closed chamber, generally constructed of steel, and divided into several compartments each of which may be operated independently of the others. Air, heated over suitably located steam pipes to the required temperature, is forced through the various compartments by means of fans operated by power. The conditions may be varied in each compartment, as to temperature or humidity, both of which can easily be regulated, or all the compartments may be used together as one unit. The skins are hung up on rods or lines in the compartments, or on special frames for the purpose, which are then entered into the compartments and the doors shut. The dry, heated air is forced to pass over the skins, and takes up their moisture. At the further end of the drying chamber is another fan which removes the moisture-laden air after it has done its work. The drying is effected in from 6 to 24 hours, and all skins are obtained in the same condition, for the process is quite uniform and regular.

Within recent years there has been evolved a highly efficient and economical drying equipment, based on a somewhat different principle than underlies any of the foregoing methods. The conveyor type of dryer, as it is called, is admirably suited to the needs of the fur dressing and dyeing industry, and is undoubtedly superior to any of the previous systems of drying furs, in that it affords an enormous saving of space, time, labor and power, and gives greater uniformity and presents better working conditions.

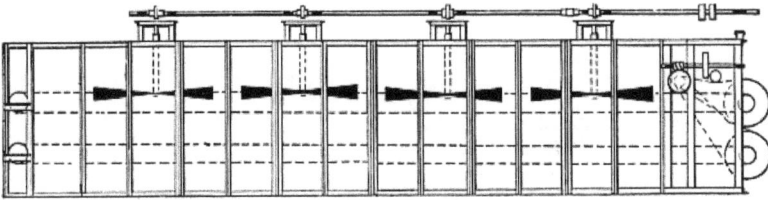

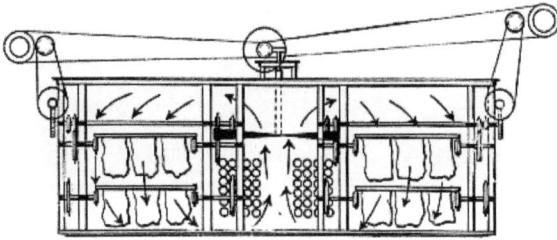

Fig. 10. Diagrammatic Views of Conveyor Dryer.

a. SIDE VIEW; *b.* END VIEW.

(*Proctor & Schwartz, Inc., Philadelphia.*)

The conveyor dryer consists essentially of a steel enclosure, through which the skins pass on horizontal conveyors. Where special insulation is necessary, asbestos panels are used to line the enclosure, making the dryer absolutely fireproof, and enabling the maximum utilization of heat. In the middle of the dryer are located the steam coils which furnish the heat, and in many instances exhaust steam can be used as the source of heat. Figure 10 shows diagrammatically the arrangement and operation of the conveyor type of dryer. The enclosure is divided into several compartments, in each of which a different condition of temperature and humidity is maintained, the temperature being closely and accurately regulated by an automatic control, and once the dryer has been set for any condition, all skins will be dried exactly the same, regardless of weather or season.

Fig. 11. Conveyor Dryer.

(Proctor & Schwartz, Inc., Philadelphia.)

The skins to be dried are placed on poles which in turn are set on the horizontal conveyors as in Fig. 11. As the skins pass through the compartments, large volumes of air, heated to the required temperature over the steam coils, are circulated among the skins by means of the fans. Exhaust fans, properly placed, remove a certain quantity of moisture-laden air when it has accomplished its full measure of work. When the skins on the conveyors have passed the full length of the dryer, they are entirely dry, and are then removed from the poles. (Fig. 12). The time required for drying varies according to the nature of the fur from 1–2 hours to 6–8 hours. In tests made to determine the relative efficiency of the conveyor type of dryer as against the old "loft" method, it was found that there was a saving of over 50% in power, and of 85% in floor space, as well as a great saving of labor, when the conveyor system was used, the number of skins dried in a given period of time being the same in both cases. The advantages of the new method are easily apparent, and the saving is sufficiently great with large lots of furs, to make an appreciable difference in the final cost of dressing.

If the skins have been dried by a modern drying system they all come out in a uniform condition, and are ready to go on immediately to the next operation. If, however, a form of the "loft" method of drying has been used, it is customary to subject the skins to an additional process. The dried pelts are put in drums with damp sawdust, and drummed for a short time in order to get them into the proper condition. The drumming is essential for the purpose of equalizing the condition of the pelts, some being drier than others, and as a consequence of the contact with the moist sawdust, they are all brought to the same degree of dryness. As a result of this operation also, the skins become considerably softened.

Fig. 12. Delivery End of Conveyor Dryer.

(*Proctor & Schwartz, Inc., Philadelphia.*)

Then if the pelts have not been previously oiled during the tanning process, or prior to the drying, they receive this treatment now. The oil or fat is applied to the leather side of the furs, which are then placed in the tramping machine for a short time in order to cause the oil to be forced into the skin. The fibres of the corium thus become coated with a thin layer of fatty material, which contributes greatly to the softness and flexibility of the pelt, and increases its resistance to the action of water, and also, in certain instances a partial chamois tan is produced, thereby improving the quality of the leather.

Fig. 13. Stretching Machine for Cased Skins.

(Reliable Machine Works, Evergreen, L. I.)

The skins are now returned to the work bench, and subjected to the stretching or "staking" process. This consists in drawing the skin in all directions over the edge of a dull blade, which is usually fixed upright in a post with the edge up. Or, the stretching may be done on the fleshing bench, substituting a dull blade for the fleshing knife. Recently staking machines are being used in the larger establishments, the work being done much more quickly and efficiently. As a result of this operation, the leather becomes very soft and flexible, every bit of hardness and stiffness being eliminated, and the skins receive their maximum stretch, thereby giving the greatest possible surface to the pelage. This not only helps to bring out the beauty of the hair, but is also a decided advantage from the economic point of view, as a considerable saving of material is effected in this way, sometimes even to the extent of twenty-five per cent. Cased skins are stretched in a somewhat different manner, by means of stretching irons. These consist of two long iron rods joined by a pivot at one end. The skins are slipped on to the irons, which are then spread apart, and in this way the skins are stretched and softened. A machine which does this work very efficiently is shown in Fig. 13. The skin is drawn onto the stretching arms, in this case made of bronze, which are then forced apart by pressing on a pedal. When properly stretched to the maximum width in all directions possible, and thus thoroughly softened, the skin can easily be reversed, that is, turned hair-side out. As many as 6000 skins can be stretched, or 4000 to 5000 skins stretched and reversed by one man in one day on such a machine.

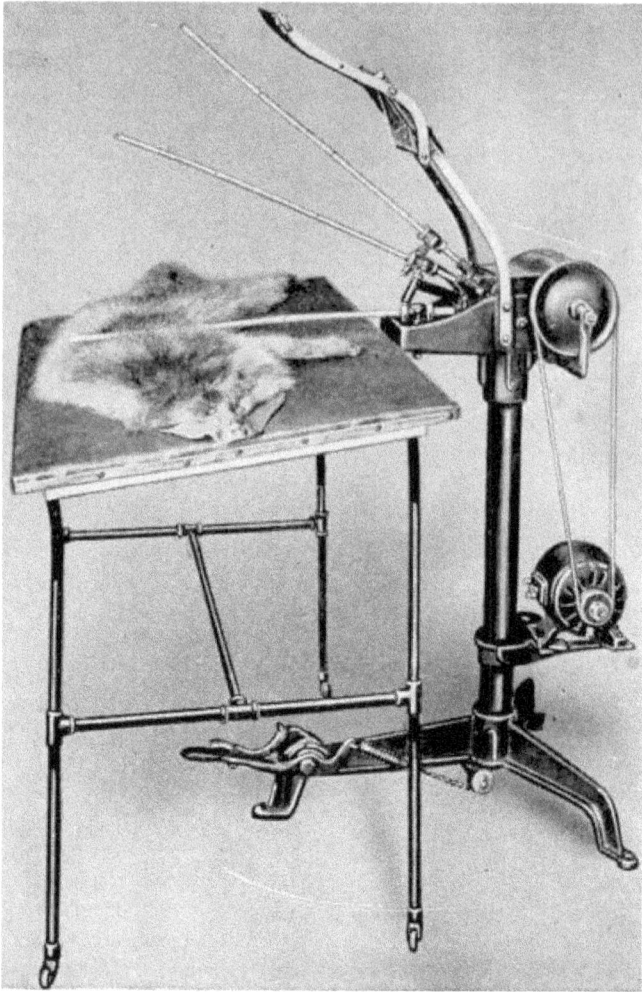

Fig. 14. Fur Beating Machine.

(*S. M. Jacoby Co., New York.*)

The pelts are then combed and beaten. In smaller plants these operations are done by hand, but suitable machines are being employed. In order to straighten out the hair, it is combed or brushed. Then in order to loosen up the hair, and to cause it to display its fullness, the furs are beaten. This process is also done by hand in some establishments, but up-to-date places use mechanical devices for this purpose. A type of machine which has proven very successful, and is enjoying considerable popularity is shown in Fig. 14. These machines are also made with special suction attachments which remove all dust as it comes out of the beaten skin, thereby making this formerly unhealthful operation thoroughly sanitary and hygienic.

The final process is drum-cleaning. This operation is intended specifically for the benefit of the hair part of the fur, and is very important inasmuch as the attractive appearance of the fur depends largely upon it. The drum, such as is shown in Fig. 15 is generally made of wood, or sometimes of wood covered with galvanized iron. The skins together with fine hardwood sawdust are tumbled for 2 to 4 hours, or sometimes longer. Occasionally a little asbestos or soapstone is added to the sawdust; for white, or very light-colored skins, gypsum or white sand is used, either alone, or in admixture with the sawdust; and for darker skins, graphite or fine charcoal is sometimes added in small quantities. The drum-cleaning process polishes the hair, giving it its full gloss and lustre, and at the same time absorbing any oil or other undesirable matter which may be adhering to the hair as a result of the washing and tanning processes. Any soap, or traces of mordant are wiped off and so removed, and by using heated sawdust, or heating the drum while rotating, the fur acquires a fullness and play of the hair which are great desiderata in furs. The sawdust must then be shaken out of the furs. This is done by cageing. In some instances, the drum itself can be converted into a cage, by replacing the solid door with one made of a wire screen. (Fig. 16.) Usually, however, the skins are removed from the drum and put in a separate cage, which is built like the drum, but has a wire net all around it, through which the sawdust falls, while the skins are held back. The cages are generally enclosed in compartments in order to prevent the sawdust from flying about and forming a dust which would be injurious to the health of the workers. In large establishments, the drum-cleaning machinery occupies a large section of the plant, many drums and cages being used, and special arrangements being made to take care of the sawdust which can be used over again several times, until it becomes quite dirty.

Fig. 15. Drum. (Combination Drum and Cage as a Drum.)

(F. Blattner, Brooklyn, New York.)

Fig. 16. Cage. (Combination Drum and Cage as a Cage.)

(*F. Blattner, Brooklyn, New York.*)

With this operation ends the ordinary procedure of fur dressing. But there are several additional processes required in the treatment of certain furs, which are generally undertaken by the dresser, and chief among these are shearing and unhairing. Sometimes this work is done in separate establishments organized solely for this business. Certain kinds of furs, among them being seal, beaver and nutria, possess top-hair which may detract from the beauty of the fur, the true attractiveness being in the fur-hair. The top-hairs are therefore removed, and for this purpose machines are now being used. Formerly this work was all done by hand, and on the more expensive furs like seal and beaver, unhairing is now done on a machine operated by hand. The principle of the process is as follows: The skins are placed on a platform and the hair blown apart by means of a bellows. The stiff top-hairs remain standing up, and sharp knives are brought down mechanically to the desired depth, and the hair is cut off at that point. The skin is then moved forward a short distance, and the process repeated until all the top-hairs have thus been cut out. With muskrats, or other pelts which do not require such very careful attention, the whole process is done automatically on a machine. The fur-hair is brushed apart by means of brushes and a comb, and at regular intervals, sharp knives cut off the top-hairs. Several hundred skins can be unhaired in a day on such a machine requiring the attention of only one man. A machine for unhairing skins is shown in Fig. 17.

Fig. 17. Unhairing Machine.

(Seneca Machine & Tool Co., Inc., Brooklyn, N. Y.)

With other furs, such as rabbits, hares, etc., where the trouble of unhairing would be too great commensurate with its advantages, the hair is sheared instead. The top hair is cut down to the same length as the under-hair by means of shearing machines which can be regulated to cut to any desired length of hair. A typical device for shearing furs is shown in Fig. 18.

Fig. 18. Fur-Shearing Machine.

(Seneca Machine & Tool Co., Inc., Brooklyn, N. Y.)

CHAPTER VII
WATER IN FUR DRESSING AND DYEING

The assertion has often been made, although its absurdity is now quite generally realized, that the success of the European fur dressers and dyers, particularly in Leipzig, is due to the peculiar nature of the water used, which is supposed to be especially suited for their needs. The achievements in this country in the fur dressing and dyeing industry during the past few years are ample and sufficient answers to the claim of foreign superiority in this field no matter what reason may be given, and particularly when the quality of the water used is advanced as a leading argument. For the water employed by the establishments in and about New York, as well as in other sections of the country is surely not the same as the water of Leipzig, yet the work done here is in every respect the equal of, if not better than the foreign products.

It is interesting to note that similar rumors were current here in the early period of the development of the American coal-tar industry since 1914. Our efforts to establish an independent dyestuff industry were doomed to failure, according to those who circulated the stories, because we did not have the water, which they claimed was responsible for the German success. The present status of the American dye business, in its capacity satisfactorily to supply most of the needs of this country and of others as well, speaks for itself.

However, as is often the case with such erroneous assertions, there is just enough of an element of truth in the statement regarding the peculiar qualities of certain kinds of water, to make the matter worthy of consideration. Water is certainly a factor of great importance in fur dressing and dyeing, and it is not every sort of water that is suitable for use. This fact was recognized by the early masters of the art, for they invariably used rainwater as the medium for their tanning and dyeing materials, and their choice must be regarded as an exceedingly wise one. While the necessity for giving consideration to the quality of the water for fur dressing purposes is great, it is in fur dyeing that the effects of using the wrong water are largely evident, and so extra care must be exercised in the selection of water for this purpose.

The essential requirements for a water suitable for the needs of the fur dressing and dyeing industry, are: first, a sufficient, constant and uniform supply; and second, the absence of certain deleterious ingredients. Chemically pure water is simply the product of the combination of two parts by volume of hydrogen with one part by volume of oxygen. Such water can only be made in the laboratory, and is of no importance in industry. For practical purposes, distilled water may be regarded as the standard of pure water. Here, too, the cost and trouble involved in the production of distilled water on a large scale is warranted only in a certain few industrial operations. A natural source of water which in its character most nearly approaches distilled water is rain. In fact, rain-water is a distilled water, for the sun's heat vaporizes the water from the surface of the earth forming clouds, which on cooling, are condensed and come down as rain. Rain-water is usually regarded as the purest form of natural water. Exclusive of the first rain after a dry period, rain-water is quite free of impurities, except possibly for a small percentage of dissolved atmospheric gases, which are practically harmless, and which can usually be readily eliminated by heating the water. Moreover, rain-water is quite uniform in its composition throughout the year in the same locality, and it possesses all the desirable qualities of a water suited for fur dressing and dyeing purposes. Formerly when the quantity of water used in the industry was comparatively small, the supply from rain was sufficient to meet all the requirements. But now, when tremendous quantities of water are used constantly, rain-water is no longer a feasible source, and other supplies must be utilized, although in a sense, all water may be traced to rain-water as its origin.

When rain-water falls on the earth it either sinks into the ground until it reaches an impervious layer, where it collects as a subterranean pool, forming a well, or continues to flow underground until it finally emerges at the surface as a spring; or on the other hand the rain-water may sink but a short distance below the surface, draining off as ponds, lakes or rivers. In the first case the water is called ground water, in the latter it is known as surface water. Ground water usually contains metallic salts in solution, and relatively little suspended matter. If the water has percolated through igneous rocks, like granite, it may be quite free even of dissolved salts, and such water is considered "soft." If, however, the rocky formations over which, or through which, the water has passed contain limestone or sandstone, or the like, salts of calcium and magnesium will be dissolved by the water. The presence of the lime and magnesia salts, as well as salts of aluminum and iron, in the water, causes it to be what is termed "hard." Surface water is more likely to contain suspended matter, with very little of dissolved substances. Suspended matter, like mud, contains much objectionable matter such as putrefactive organisms and iron, but most of these materials can be removed by filtration or sedimentation, and seldom cause any difficulties.

Hardness in water is generally the chief source of trouble when the water is at fault. Hardness may be of two kinds, either permanent, or temporary, or sometimes both are found together. Water which is permanently hard usually contains the lime and magnesia combined as sulphates. Temporary hardness, on the other hand, is due to the presence of lime and magnesia in the form of bicarbonates, the carbon dioxide contained in the water having dissolved the practically insoluble carbonates:

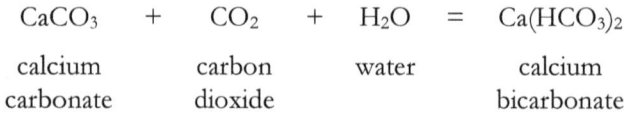

$$CaCO_3 \quad + \quad CO_2 \quad + \quad H_2O \quad = \quad Ca(HCO_3)_2$$

calcium   carbon   water   calcium
carbonate   dioxide   bicarbonate

Temporary hardness can be eliminated by heating the water, the carbon dioxide being expelled and the carbonates of lime and magnesia being precipitated and then filtered off. Both permanently and temporarily hard waters can be softened by the addition of the proper chemical, such as an alkaline carbonate like sodium carbonate. This precipitates insoluble carbonates of the lime, magnesia, iron and aluminum, leaving a harmless salt of sodium in solution in the water. The sludge is allowed to settle in tanks before the water is used.

In fur dressing and dyeing, water is employed for soaking and washing the skins, dissolving chemicals, extracts and dye materials, and also for steam boilers. A small amount of hardness in the water is not harmful, and up to 10 parts of solid matter per 100,000, may be disregarded. Permanent hardness is particularly objectionable in water for boiler purposes, as it forms scale. The effect of the impurities of the water depends on the nature of the chemicals and dyes used. Where acids are used in solution compounds of magnesium, lime and aluminum will generally not interfere. Hard water must not be used for soap solutions, as sticky insoluble precipitates are formed with the soap by the metals, this compound adhering to the hair, and being difficult to remove, will cause considerable trouble in subsequent dyeing. An appreciable loss of soap also results, as one part of lime, calculated as carbonate will render useless twelve parts of soap. In tanning or mordanting, where salts of tin, aluminum or iron are employed, hard water should not be used, as lime and magnesia will form precipitates with them. Bichromates will be reduced to neutral salts, and cream of tartar will also be neutralized. With dyes also, hard water has a deleterious effect. Basic dyes are precipitated by this kind of water, rendering part of the dye useless, and also causing uneven and streaky dyeings. Sometimes the shades of the dyeings are modified or unfavorably affected. Considerable quantities of lime and magnesia in the water will cause duller shades with logwood and fustic dyeings. The presence of iron, even in very slight quantities generally alters the shade, darkening and dulling the color.

These facts were apparently all recognized and understood by the fur dressers and dyers of an earlier period, for instead of utilizing the water of lakes and streams near at hand, which afforded a more constant supply, but which contained harmful impurities, they collected the rain-water, which was always soft. Whether they realized the nature and character of the substances that make water hard is uncertain, but they were always careful to avoid such water. At the present time establishments located in and about large cities like New York, where the majority of American fur dressing and dyeing plants are situated, have no trouble about the water. The cities supply water which is soft, suitable alike for drinking and industrial purposes. Other plants, not so fortunately situated, often have to employ chemical means to treat the water so as to make it suitable for use.

CHAPTER VIII
FUR DYEING
INTRODUCTORY AND HISTORICAL

In discussing fur dyeing, the question naturally arises, "Why dye furs at all? Are not furs most attractive in their natural colors, and therefore more desirable than those which acquire their color through the artifices of man?" The answer cannot be given simply. Natural furs of the more valuable kinds are indeed above comparison with the majority of dyed furs. Yet there are several reasons which fully justify and explain the need for fur dyeing, for at the present time, this branch of the fur industry is almost as important and indispensable as the dressing of furs.

The first application of dyeing to furs, had for its purpose the improvement of skins which were poor or faulty in color; or rather, the object was to hide such defects. As nearly as can be ascertained, this practise was instituted at some time during or before the fourteenth century, for fur dyeing seems to have been common during that period, as is apparent from the verses of a well-known German satirist, Sebastian Brant, who lived in the latter part of the fourteenth century:

"Man kann jetzt alles Pelzwerk färben,

Und tut es auf das schlechste gerben."

However, at a later period, there was a general condemnation of the dyeing of furs, and among the list of members of the furrier's guilds, none can be found who are described as dyers. There is a record of a decree issued by a prince in a German city in the sixteenth century, prohibiting the practise of fur dyeing. Inasmuch as furs were worn only by the nobility and certain other privileged classes, and also were very costly, there was great profit to be had by dyeing inferior skins so as to disguise the poor color, and then selling such furs at the price of superior quality skins. This was undoubtedly the reason for the prohibitory decree, but there were some who continued to practise the forbidden art in secret, using secluded and out-of-the-way places for their workshops, and mixing their carefully-guarded recipes with as much mystery as the witches did their magic potions. These circumstances probably account for the great amount of mystery which has been, and still is to a considerable degree, attached to fur dyeing, and also explains the opprobrium and distrust with which fur dyers were formerly regarded.

Even at the present time, dyeing is often employed to improve furs which are faulty in color. It frequently happens, that in a lot of skins there are some which are considerably off shade, or in which the color is such as to appreciably reduce their value below the average, the hair being usually too light a shade, or of uneven coloring. By carefully dyeing these skins of inferior color, they can be made to match very closely the best colored skins of the particular lot of furs, and consequently increase their value. With most of the cheaper kinds of furs, the trouble and cost of improvement by dyeing would not be worth while today; but with some of the more valuable furs, and especially such as are very highly prized, like the Russian sable, or marten, or chinchilla, the darkening of light skins by the skillful application of fast dyes to the extreme tips of the hair, will increase their value sufficiently to warrant the expense. This dyeing or "blending" as it is called in such cases, is done in such a clever and artistic manner that only experts can distinguish them from the natural. Dyeing used for such purposes is not objectionable, provided the skins are sold as dyed or "blended."

There are certain kinds of furs, such as the various lambs, Persian, Astrachan, Caracul, etc., which are never used in their natural color, because it is usually of a rusty brownish-black. These are furs possessing valuable qualities otherwise, so they are dyed a pretty shade of black, which brings out the beauty of the fur to the fullest extent. Sealskins are also dyed always. Formerly they were dyed a deep, rich dark brown, resembling the finest shades of the natural color, but now the seals are dyed black with a brownish undertone, a color quite different from the natural. While these two instances cannot be said to be cases of dyeing to disguise faulty color, they are examples of improvement of color by dyeing.

Closely associated with the use of dyes to increase the value of a fur by improving its color, is the dyeing of skins of a certain lot of furs to produce a uniform shade, thereby facilitating or to a considerable degree eliminating the task of matching the skins by the furrier. This is usually done only on skins which are quite small, of which a great many are needed in the manufacture of fur garments, because the matching of several hundred skins would entail too much time and labor commensurate with the value of the fur. The most notable instance of the use of dyes to produce a uniform shade on furs is the case of the moleskin. Occasionally, furs are dyed after being made into garments, by careful application of dyes, in order to obtain certain harmonious effects, such as uniformity of stripe, or to produce a desired gradation of shade among the different skins comprising the garment.

Not infrequently, the great variety of shades and color schemes which Nature provides in the different furs, becomes insufficient to satisfy the desire of the fur-wearing public for something new. The whims of fashion always require some novel effect, even though it be for only one season. To meet this demand for novelty, fantasy or mode shades are produced on suitable furs,—colors which do not imitate those of any animal at all, but which, nevertheless, strike the popular fancy. It often happens that such a color becomes quite popular, and enjoys a considerable vogue, to the great profit of those who introduced the particular color effect. The best ones, however, meet with only a comparatively short-lived demand, being soon superseded by different color novelties.

The basis, though, of the greatest proportion of fur dyeing at the present time, is the imitation of the more valuable furs on cheaper or inferior skins. With the gradual popularization of furs as wearing apparel since the beginning of the last century, the demand for furs of all kinds has increased enormously. The supply of furs, on the other hand, and especially of the rarer kinds, has had difficulty in keeping pace with the requirements, and as a result there is a shortage. A very effective means of relieving this shortage, to a great degree, at any rate, is the dyeing of imitations of the scarcer furs on cheaper skins. There are many animals among the more common, and more easily obtainable ones, whose skins are admirably suited as the basis of imitations of the more costly furs. Some of the furs which are adapted for purposes of dyeing imitations are marmot, red fox, rabbit, hare, muskrat, squirrel, opossum, raccoon, and many others, and the imitations made are those of mink, sable, marten, skunk, seal, chinchilla, etc., and indeed, there are very few valuable furs, which have not been dyed in imitation on cheaper pelts. On account of the general mystery which formerly surrounded fur dyeing establishments, and which has persisted to this day, although to a lesser degree, many peculiar notions were held, even by those in the fur trade, concerning the production of imitations. The idea that in order to "make" a certain fur out of a cheaper skin, it was necessary to use the blood of the animal imitated, is typical of the conceptions of fur dyeing held not so long ago. To-day, while the knowledge generally possessed about this branch of the fur industry is meagre and vague, the air of mystery and secrecy has become somewhat clarified, and such ideas as are current about fur dyeing are more rational than formerly.

The dyeing of imitations is quite an artistic kind of work, and indeed fur dyeing ought to be classed among the finest of industrial arts. Some of the reproductions achieved by dyers on a commercial scale are truly admirable. The possibility of imitating the finer furs on cheaper skins naturally led to abuse, the dyed furs being passed off frequently on the unsuspecting and uninformed buyer as the genuine original. In fact, this practise became so flagrant that in England laws were enacted to remedy the evil. At the present time, dyed furs are all sold as such, although there always may be some unscrupulous merchants who seek to profit by deception. Some of the imitations and the names of the furs for which they were sold, are as follows:

Muskrat, dyed and plucked	sold as seal
Nutria, plucked and dyed	sold as seal
Nutria, plucked and natural	sold as beaver
Rabbit, sheared and dyed	sold as seal or electric seal
Otter, plucked and dyed	sold as seal
Marmot, dyed	sold as mink or sable
Fitch, dyed	sold as sable
Rabbit, dyed	sold as sable
Rabbit, dyed and sheared	sold as beaver
Muskrat, dyed	sold as mink or sable
Hare, dyed	sold as sable, fox, or lynx
Wallaby, dyed	sold as skunk
White rabbit, natural	sold as ermine
White rabbit, dyed	sold as chinchilla
White hare, dyed or natural	sold as foxes, etc.
Goat, dyed	sold as bear, leopard, etc.

This list serves to indicate but a few of the great number of possibilities which are available for the fur dyer to produce imitations of the better classes of furs. Needless to say, these imitations cannot, as a general rule, equal the originals, because while the color is one of the most important features in judging the fur, the nature of the hair, gloss, waviness, thickness, and also the durability are essential considerations, and it is only in certain instances that skins used for imitations approach the originals in these respects. However, for the purposes and desires of the majority of people who wear furs, the imitations are deemed quite satisfactory, and they also have the advantage of being cheaper than the natural originals.

For whichever reason furs are dyed, there is no doubt that the art of fur dyeing is one of the most difficult kinds of application of dye materials. In the dyeing of the various textiles, either as skein or woven fabric, the material is of a uniform nature, and therefore the dye is absorbed evenly by the fibres. Moreover, textiles are dyed at, or near the boil, the dyestuff being more uniformly and permanently taken up from solution by the fibre at elevated temperatures.

How different is the case with furs! Far from being homogeneous, furs present the greatest possible diversity of fibres to be dyed. As already noted elsewhere, fur consists of two principal parts, the hair and the leather, differing widely in their actions toward dyes. As a general rule, the leather absorbs dyestuffs much more readily than the pelage, and inasmuch as fur dyeing is intended mainly and primarily to apply to the hair, there is usually an appreciable loss of dye material due to its being absorbed by the leather, and thereby rendered unavailable for dyeing the hair. This fact must be taken into account in the dyeing of furs, and the methods must be adapted accordingly.

With reference to the hair itself, not only has each class of furs hair of a different kind, but even in the same group there is always a considerable divergence in the properties of the hair. The fur-hair, being more or less of a woolly nature, takes up the dye with comparative ease, while the top-hair is quite resistant to the action of all dye materials. As pointed out in the discussion of the nature of fur, on different parts of the same pelt the hair varies in its capacity for absorbing coloring matters. The color of the hair, also frequently presents a great variety throughout the skin, both in fur-hair and top-hair. Yet with all this lack of uniformity and homogeneity, the dyed fur must be of an even color, closely approaching the natural, gently graded and without any harsh or unduly contrasted effects. The natural gloss of the hair, one of the most valuable qualities of the fur, must be preserved. This is by no means a simple matter, for the luster is affected by dyes and chemicals with comparative ease, and especially careful treatment is necessary to prevent any diminution of the gloss.

When the leather part of the fur is exposed to solutions of a temperature exceeding 40°–50° centigrade, it soon shrivels up or shrinks, and on drying the pelt, becomes hard and brittle, and therefore quite useless. Methods of fur dyeing have to take into consideration this fact, and the temperature of the dyebath must not be greater than 35°–40° centigrade. To be sure, certain dressings make furs capable of withstanding much higher temperatures, but their applicability is not universal, being suited only for a very limited special class of dyestuffs. (V. Fur Dressing). The necessity for employing comparatively low temperatures, coupled with the great resistance of the hair to the absorption of dye, even at much higher temperatures, makes fur-dyeing a very difficult operation indeed. Another obstacle which must be surmounted, is the possibility of extraction by the dye solution, of those materials, chemical or otherwise, which are contained in the leather, and which are the basis of its permanence, softness and flexibility. For in the majority of dressing processes, the action of the ingredients is a preservative one, and when these are wholly or partially removed from the leather during the dyeing, it becomes, on drying, hard and horny, like the original undressed pelt. In cases where furs are to be dyed, special dye-resisting dressings must be used, or the dyed skins must receive an additional dressing before drying.

Dyeings on furs, to have any value, must possess great fastness to light, rubbing and wear, and must not change color in time, either when the furs are stored, or when made up into garments. The necessity for fur dyeings to have these properties, together with the difficulties outlined above, has greatly limited the field of available dyeing materials, as well as the methods of application. These will now be taken up in detail.

CHAPTER IX
FUR DYEING
GENERAL METHODS

Before the furs can be dyed, they have to undergo certain preparatory processes: first, killing, which renders the hair more susceptible to the absorption of the dye; and second, mordanting, which consists in treating the killed fur with chemicals which help the dye to be fixed on the hair. Then the skins are ready to be dyed.

There are two principal methods by which dyes are applied to furs in practise: the brush process, whereby only the tips or the upper part of the hair are colored; and the dip process, whereby the entire fur, including the leather is dyed. All other procedures in fur dyeing are modifications or combinations of these two. Killing solutions and mordanting solutions are also applied by one of these methods, usually the dip process, although very frequently combinations of the brush and dip methods are used.

Chronologically the brush method of dyeing came first. The early masters of the art were extremely fearful about employing any means by which there was a possibility of the leather being in any way affected. They naturally had to devise such methods as would give the desired effect in a satisfactory manner, and as would be confined solely to the hair part of the fur, leaving the leather untouched. By applying the dye or other material to be used, in the form of a paste with a brush, the upper portion of the hair only was treated. For different kinds of furs different sorts of brushes were used, and the depth to which the hair was colored could be controlled by skillful manipulation of the brushes. It was frequently necessary to give a ground color to the hair, the lower part being dyed a different shade from the tips. This was accomplished by spreading the dye paste over the hair with a broad brush, and then beating the color in with a specially adapted beating brush. With larger furs, two skins were placed hair to hair after the dye had been brushed on, and the color forced to the bottom of the hair by a workman tramping on the skins. The dyeing of seal was a typical illustration of these procedures. First the tips of the hair were dyed. The color was brushed on, allowed to dry, then the excess beaten out with rods. These operations were repeated until the proper depth of shade was obtained, often as many as a dozen or more applications of the dye being necessary. Then the base color was spread over the hair, and beaten or tramped in until the lower parts of the hair were penetrated. This process also required drying and beating out of the excess dye, as well as numerous applications of the dye to impart the desired color to the hair. Prior to the dyeing, the furs were killed, by brushing on a paste containing the essential ingredients, drying and beating and brushing the fur, just the same as in dyeing. It will be readily seen that such methods were exceedingly laborious, and in some cases the dyeing took many weeks, and even months.

It was quite a step forward when a certain fur dyer, possessing a little more courage, or perhaps, experimenting spirit than the others, attempted to dye furs by dipping them entirely into a bath containing a solution of the dye instead of applying a paste as formerly. The advantages to be gained by such a method of dyeing were many. A large number of skins could be treated thus at one time, and this was a very important consideration in view of the great increase in the demand for dyed furs. By allowing the furs to remain in the dye solution until the proper shade was obtained, the time and labor of applying many coats of dye by brush was considerably reduced, and in addition, there was a greater probability of the products coming out all alike, uniformly dyed. The results as far as the hair was concerned, were indeed highly gratifying, but the condition of the leather after dyeing was not so encouraging. This difficulty has to a considerable degree been overcome, although there are frequent instances of the leather being affected by the dyeing process even with modern methods. However, the remedy in such cases, or rather the preventative is the proper dressing of the skins prior to the dyeing. The dip method of dyeing has acquired great importance, and is being employed in dyeing operations involving the handling of millions of skins annually. In certain instances, nevertheless, the brush method is of prime significance as in the dyeing of seal, and seal imitations on muskrat and coney, enormous quantities of furs being dyed in this fashion. In the majority of imitations dyed, both the brush and the dip methods must be used.

Figure 19 illustrates the various types of brushes which are used at the present time for the application of the dye by the brush method. Each brush has a specific purpose and use. The procedure in brush dyeing is somewhat as follows. The skins, after being properly treated, that is, killed, and mordanted, are placed on a table, or work-bench, hair-side up. Then by means of a brush which is adapted to the nature and requirements of the particular fur, the solution is brushed on in the direction of the fall of the hair, occasionally beating gently with the brush so as to cause the dye to penetrate to the desired depth. Considerable skill and care must be exercised in this operation as it is rather easy to force the dye down further than is wanted, and in some cases the leather or the roots of the hair may be affected. The skin having received its coat of dye, is then dried and finished, if no other dyeing processes are to be applied. Frequently, with certain types of dyes, several applications of color are necessary, and these are brushed on as the first one, drying each time. Then, on the other hand, the skin may receive a dyeing in the bath by dipping, and for this also, the fur is first dried after the brush dyeing.

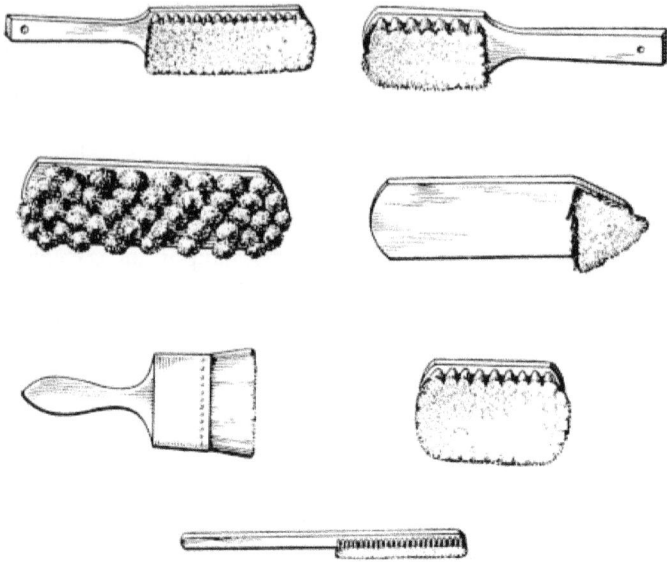

Fig. 19. Brushes Used in Fur Dyeing By the Brush Method.

Quite recently, owing to the great quantities of furs which are being dyed as seal imitations, chiefly by the brush method, although the dip method is used in conjunction with it, machines have been invented to replace the hand brush, and the dye is now applied mechanically. Machines for this purpose are by no means new, there being records of inventions almost a score of years past, but they did not achieve much success. Brush-dyeing machines, to be efficient, must be designed to suit the needs of the particular type of fur to be dyed, otherwise there will be a great lack of uniformity in the dyed skins, a condition which cannot occur when the dye is brushed on by hand brushes. Figure 20A and B shows diagrammatically, machines invented within the past few years, which are used to dye mechanically furs by the brush process.

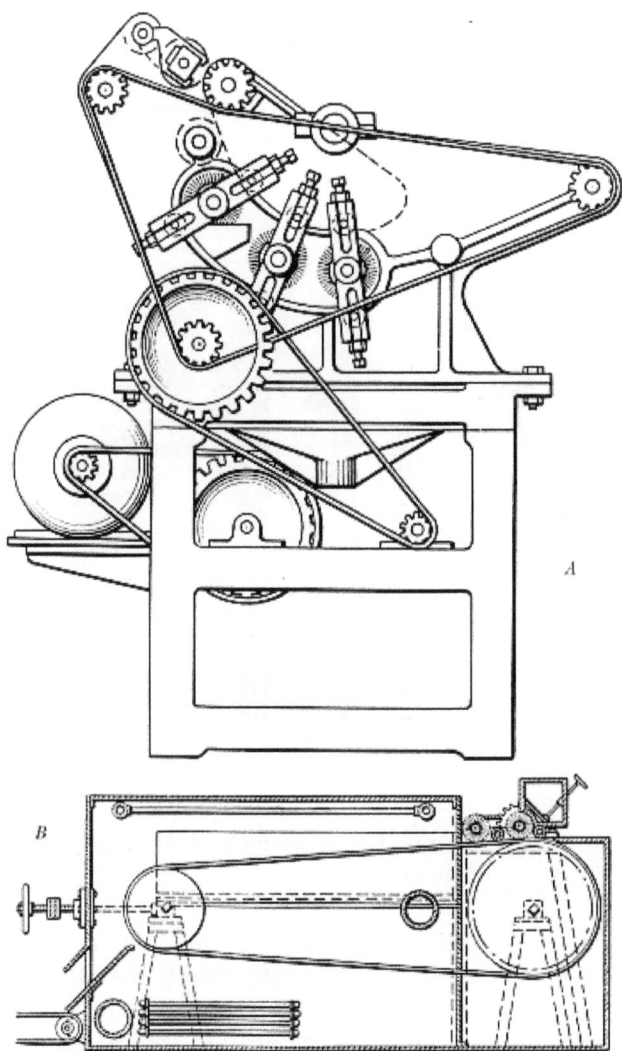

Fig. 20. Types of Machines for Dyeing Furs By the Brush Method.

A. (U. S. Patent 1,225,447.) *B*. (U. S. Patent 1,343,355.)

Fig. 21. Drum For Working With Liquids.

(Turner Tanning Machinery Co., Peabody, Mass.)

For the dipping process, the dye solution is prepared in vats, or liquid-tight drums, or in some instances in paddle arrangements. The skins are placed in the dye-bath, and the dyeing operation proceeds without any difficulty. After the proper shade is obtained, the furs are removed, washed free of excess dye, dried and finished. The dipping method is employed where a single shade is to be dyed on the fur, as the production of blacks on lambs. But in most cases, the dyeing in the bath is supplemented by the application of a coat of dye by the brush to the upper part of the hair, the color being usually a darker shade than the ground dyeing. Thus, for example, in the dyeing of imitation sable on kolinsky or a similar fur, the skins are first dyed the relatively light color of the under-hair by the dip process, then the dark stripe effect is brushed on.

The blending of sables, martens, chinchillas or other rare furs, is not done in the same manner as with other furs, because each skin requires individual attention and a long and careful treatment. The dye solution is applied by means of very fine brushes or sometimes feathers, to the extreme tips of the hair, until the proper degree of color intensity is obtained. The time, labor, and skill necessary for this sort of work are warranted only in the case of the highest-priced furs, and the blendings are so excellent as to defy detection, except by experts.

Fig. 22. Device For Conveying Skins.

(*Turner Tanning Machinery Co., Peabody, Mass.*)

After the furs have gone through all the operations required by the processes of killing, mordanting, dyeing and washing, they are ready to be dried and finished. The procedure is quite similar to that employed in fur dressing. Sometimes the leather side of the skins is brushed with a strong salt solution before drying, in order to replace some of the salt which was extracted during the dyeing processes. In other instances, a light coat of some oily substance is brushed on, to render the leather soft and flexible after drying, where there is a possibility of the skins turning out otherwise. Great care must be exercised in the handling of the dyed skins to avoid the formation of stains or spots on the hair, which might ruin the dyeing. As little handling of the furs as is feasible will reduce any trouble from this source. In conveying the wet skins from one part of the plant to another it is desirable to use a device such as is shown in Fig. 22. For drying, the same machines as described under Fur Dressing can be used, and similar care must be taken to avoid overheating or irregularity of drying. Drum-cleaning constitutes a very important operation in the finishing of the skins, the hair receiving a polish, and the full lustre and brilliancy of the dye being thereby brought out. Then after caging to remove the sawdust or sand, the skins are passed over the staking knife, or are treated in a machine suited for the purpose, to stretch them and to render them thoroughly soft and flexible. And therewith is concluded the work of the fur dyer proper, and the skins are ready to return to the furrier, in whose hands they undergo the metamorphosis into the fur garments to be worn chiefly by the feminine portion of humanity.

CHAPTER X
FUR DYEING
"KILLING" THE FURS

If dressed furs are treated with a paste or solution of a dye properly prepared, and at the right temperature, the hair will show very little tendency to absorb the coloring matter. Even after prolonged treatment with the dye, only a small amount will be taken up by the hair, and in a very irregular fashion. Soft, woolly hair, like that of lambs and goats will be colored more easily than that of furs with harder hair, and the under-hair of a fur will generally have a greater affinity for the dye than the harder and stiffer top-hair. Moreover, in some parts of the same fur, the hair will absorb more color than in other parts. In other words, the hair of furs resists the action of dye materials to a greater or less degree, depending upon the character of the fur, and also upon the part of the pelt. In order to overcome this resistance of the hair, and to render it uniformly receptive to the coloring substances, the furs are treated with certain chemical agents, the process being known technically as "killing."

The origin of the term is obscure, but it is interesting to note that in the fur dyeing countries other than the United States and England, the corresponding expression is used: in Germany, "töten," and in France "tuer." The explanation of the process is as follows: The surface of the hair is covered with a fine coat of fatty material which renders the hair more or less impervious to dye solutions and solutions of other substances which may be used for dyeing purposes. This fatty coating of the hair cannot be removed by mechanical means, otherwise the hair would have been freed of it during the dressing operations. Chemical solvents must therefore be resorted to, and naturally alkaline materials are used, these being usually cheapest and also most effective in their dissolving action on fatty substances. Alcohol, ether, benzine, and other similar liquids also serve as killing agents on furs, since they too, are fat solvents. In all these cases, the fatty substance on the hair is dissolved away, and the protective coat which previously rendered the hair impervious to the dye, is now removed. There are certain chemicals however, which normally do not dissolve substances of a fatty nature, but are strongly oxidizing, such as peroxide of hydrogen, hypochlorites, permanganates, perborates, nitric acid, etc., and exert a killing action when they are applied to the hair, in that the hair is made capable of taking up the dye from its solutions. In this case the killing can hardly be said to be due to a degreasing process. The fact that killing can be brought about with other substances than alkalies or fat solvents, has led to the belief on the part of some investigators in this field that killing is more than a degreasing operation, although the removal of the fatty material of the hair undoubtedly takes place. Some authorities consider that the killing process changes the pigment of the hair, which thereby becomes more receptive to the dye. It is quite possible that some such change in the structure of the hair fibre does take place, the surface of the hair becoming slightly roughened, and therefore more capable of fixing the coloring matter. The question is still an open one, and since no conclusive researches have been made as yet, it will be assumed that killing is simply a degreasing process, inasmuch as the modern practise is based on this supposition, and very satisfactory results are obtained.

An account of the historical development of the killing process brings out many interesting and enlightening facts, so it will be given here briefly. One of the first substances used for killing, or degreasing the hair of furs, was decomposing urine. Urine contains about 2% of urea which gradually changes to salts of ammonia, and in the presence of the air, largely to ammonium carbonate. This substance has a weak alkaline action, but sufficiently effective to be used for killing the hair of certain types of furs. Woolly furs, such as those derived from the various kinds of sheep and goats, were degreased with stale urine, the skins being washed in this, and then rinsed in water. The fat was emulsified by the ammonium carbonate present, and could thus be easily removed. For other furs, a stronger mixture was necessary. An example of a killing formula used on wolf, skunk and raccoon, which were to be dyed black, is the following:

  350 grams beechwood ashes

  200 grams unslaked lime

  150 grams copper vitriol

  100 grams litharge

  60 grams salammoniac

  40 grams crystallized verdigris

  3.5 liters rain water

Beechwood ashes were a very important constituent of the old killing formulas. The reason for that lies in the fact that beechwood contains a comparatively high percentage of potassium, which occurs in the ashes of the burned wood as potassium carbonate, or potash. The ashes alone were frequently used, being applied in the form of a paste, which in some instances had an advantage over a solution, in that the killing could be limited to certain parts of the skin where it was more desired than in other parts. By extracting the wood ashes with hot water, and evaporating the clear solution to dryness, potash could be obtained, which was considerably stronger than the original ashes. Next in importance for the killing was unslaked lime. This substance was also often used by itself, being first slaked with water, and using the milk of lime thus formed, after cooling. Salammoniac, although a salt, and consequently without any killing action, in contact with the beechwood ashes or the lime in solution or paste, liberated ammonia slowly, and so also acted as a degreasing agent. The other chemicals in the formula took no part in the actual killing of the hair, but acted either as mordant materials or as mineral dyes. The copper salts, in this mixture present in two forms, as sulphate in copper vitriol, and as acetate in the verdigris, were important constituents of the dye formula, being essential to the production of the proper shade. These substances properly had no place in the killing formula. The litharge, also was not a killing agent, but in the presence of the alkaline materials of the killing mixture, it gradually combined with the sulphur contained in the hair, forming lead sulphide, and thereby darkening the color of the hair. In this case, the metallic compound acted, not as a mordant, but as a mineral dye. The mixture was applied to the hair by means of a brush, the skins let lie for some time, then dried, brushed and beaten. Many applications were usually necessary to sufficiently degrease the hair. Inasmuch as the killing paste was prepared by mixing the constituents together, and then was brushed on at the comparatively low temperatures which the proper protection of the hair required, it is questionable whether some of the metal compounds were even enabled to act as described above as mordant or dye. In spite of the trouble and considerable time required in working with such a killing formula to obtain the hair in the desired condition for dyeing, the use of such a mixture nevertheless possessed the advantage that the hair was only very slowly and gradually acted upon, and so the gloss was preserved. The action of strong alkaline substances acting quickly is more or less detrimental to keeping the gloss of the hair, while the slow action of the weak alkaline paste of the old formulas, and the gradual formation of a protective metal film on the surface of the hair, rendered the hair suitably receptive to the dye which was subsequently applied, without in any measure affecting the lustre of the hair.

It would be needless to describe or discuss any more of the old killing formulas, for the principle involved was the same in all cases, there being usually a slight variation in the content of metallic salts, beechwood ashes and unslaked lime being constituents of the great majority of the mixtures used. Modern killing processes employ substances quite similar to those of the old formulas, the operations, however, being much less laborious and less time-consuming, and the cheap, pure products which chemical science has been able to develop being used in place of the crude products crudely obtained from natural sources. The chemicals used at the present time for killing furs, are chiefly ammonia, soda ash, caustic soda, and caustic lime. The choice of the killing agent depends upon the nature of the fur, the hair of some furs being sufficiently killed by treatment with weak alkalies, while in other furs the hair may require stronger treatment. The ability of the hair of a particular fur to withstand the action of the different alkaline substances must be taken into consideration, there being a great divergence in this regard among the different classes of furs. Raccoon, for example, is not appreciably affected by a solution of caustic soda of 5 degrees Beaumé, while some wolf hair cannot withstand the action of a solution of soda ash of less than 1 degree Beaumé. Frequently much stronger alkalies are necessary to kill the top-hair than the under-hair, so this accomplished by treating the skins in a solution which is suited to kill the under-hair, and subsequently the top-hair is treated with a stronger solution, this being applied by the brush method.

Uniformity of action of the killing material on all parts of the skin, and on all the skins of a given lot, is absolutely essential to obtaining satisfactory results in dyeing. And it is by no means a simple matter to get such uniformity, considering the numerous factors that must be taken into account. Any operation involving the immersion of the skins in solutions or even in water alone, has an effect on the leather side of the skin, inasmuch as some of the tanning materials may be extracted. The application of some substance of a fatty nature to a great degree prevents this, and the skin can be killed, mordanted and dyed, and then come out soft and flexible. But the great majority of substances of a fatty nature are affected by alkalies, and so when the skins are being killed, the action of the alkaline materials would be upon the fat contained in the leather as well as that upon the hair. As a result the hair may not be sufficiently killed, and so give uneven dyeings subsequently. Either a certain excess of the killing chemical must be used, and it would be very difficult to ascertain what quantity would suffice, or the killing action must be prolonged; but best of all, in oiling the skins, an inert mineral oil should be used, since it is wholly unaffected by alkalies.

Skins may be killed by the brush process or the dip process, or by both. For brush killing, the stronger alkalies like lime and caustic soda are used, the solution being applied to the top-hair with a suitable brush, and the skins allowed to remain hair to hair for the necessary length of time, after which they are treated further as skins killed by the dip process. By this latter process, the furs are immersed in a solution of the desired killing agent in a vat, or drum, or other appropriate device which will permit of uniform action of the alkali on the hair of all the skins. After remaining in the solution the required length of time, the skins are drained, and rinsed in fresh water, and then entered into a weak solution of an acid in order to neutralize any remaining alkali, it being easier to wash out acid than alkali. The furs are then washed thoroughly in clear water, preferably running water, to remove the last traces of acid. The skins are then drained and hydro-extracted, or pressed, and are then ready for the subsequent operations of mordanting and dyeing.

KILLING WITH SODA

Soda is sodium carbonate, which is produced commercially in a very pure state in several different forms, the chief being sal soda, which is crystallized sodium carbonate, containing about 37% of actual soda; and soda ash, or calcined soda, which is anhydrous sodium carbonate. The latter is the variety most commonly used.

10 grams soda ash are dissolved in

1 liter of water at 25°–30° C.

The skins are immersed for 2–3 hours, after which they are rinsed and treated with

10 grams acetic acid dissolved in

1 liter of water.

The skins are again thoroughly washed, and then hydro-extracted.

KILLING WITH LIME

Lime, calcium oxide, forms a white, amorphous, porous substance, which readily takes up water, giving calcium hydroxide, or slaked lime. Only the best grades of lime should be used, as it is very frequently contaminated with calcium carbonate and other inert materials.

10 grams of lime are dissolved in

1 liter of water.

The skins are entered, and allowed to remain for a period of time which varies according to the nature of the fur. During the killing, the solution must be agitated, in order to evenly distribute the milk of lime, which has a tendency to settle out. After rinsing, the skins are "soured," by treating with weak acetic acid solution, then thoroughly washed, and drained.

KILLING WITH CAUSTIC SODA

Caustic soda is used only on furs the hair of which is very hard and resistant to killing. Usually it is applied by the brush process, but in some instances, the dip method must be used. In order to reduce as far as possible, the action of the caustic soda on the leather, the weakest permissible solutions are used, increasing the time of treatment, if necessary. Caustic soda is a white, crystalline substance, occurring in commerce in lumps, but more conveniently in a solution of 40 degrees Beaumé, containing 35% of caustic soda. Various quantities, ranging from 4 to 25 grams of this solution per liter of water are taken, according to the character of the fur, and the skins treated for 2–3 hours, although weaker solutions may be used, and increasing the duration of the killing. By keeping the solution in motion, by means of a stirrer or any other method of agitation, the best results are obtained. After the skins are sufficiently killed, they are soured, and washed as by the other killing methods.

Where the nature of the hair of the fur is such that the top-hair and the under-hair require different killing treatments, the skins are first killed by the dip process, with an alkali suited to kill the under-hair, then a brush killing with a stronger alkali is applied to the top-hair. The subsequent treatments are the same as for usual dip-killing methods.

CHAPTER XI
FUR DYEING
MORDANTS

The hair of furs has the peculiar quality of fixing the oxides or hydroxides of certain metals from dilute solutions of their salts. Advantage is taken of this property to mordant the furs, that is, to cause a certain amount of the metallic oxide or hydroxide to be permanently absorbed by the fibres. The term mordant comes from the French word "mordre," meaning to bite, it being formerly considered that the purpose of a mordant was to attack the surface of the hair in such a way as to permit the dye to be more easily absorbed. In fact, killing mixtures, which were intended for this same object, used to contain the various chemicals which have a mordanting action, in addition to the alkaline constituents. The mordants were not applied as such, but always as killing materials. It was later realized, however, that the mordant was instrumental in the production of the color itself.

Mordanting may be considered as having a two-fold object: first, to help fix the dye on the fibre in a more permanent fashion, thus rendering the dyeings faster; and secondly, to help obtain certain shades of color, as the various mordants produce different shades with any given dye. Some classes of dyes can be applied to furs without the use of mordants, but other types are taken up only in a very loose manner, being easily washed out from the hair with water, and it is only when such dyes are brought on to the hair in the form of a metallic compound, producing what is known as a "lake," that really fast dyeings are obtained with them. The substances which are used for mordanting the hair are certain metallic compounds, but not all metallic salts which are used in dyeing are mordants. Sometimes such a compound is employed to develop the color of the dyeing by after-treatment, as in the case of after-chroming, the action of the metallic salt being directed only to the dye, and is not fixed by the fibre as a mordant must be. In order for a metallic compound to act as a true mordant, it must be fixed by the hair, and it must combine with the dye, thus forming a sort of connecting link between the dye and the hair. It is not absolutely essential that the mordant be applied first, although this is the customary and commonest practise. There are three ways by which the mordants can be fixed on the fur hair: First, by the absorption of the metallic oxide or hydroxide from a solution of the mordant prior to the dyeing; second, the mordant may be fixed on the fibre at the same time as the dye; and third, the mordant may be applied after the fur has been treated with the dye. The last two methods will be discussed in connection with the dyes, as they are special cases.

The salts of metals which are comparatively easily dissociated in water, with the formation of insoluble oxides or hydroxides, are most applicable as mordants for furs, and among them are compounds of aluminum, iron, chromium, copper and tin. The constituents of the hair seem to bring about the dissociation of the metallic salt, and the oxide or hydroxide as the case may be, is absorbed and firmly fixed by the hair. Just what the manner and nature of this fixation are, is still uncertain. It is supposed that chemical combination takes place between the hair and the metal. The course of this process may, as far as is known, be described as follows, taking, for example, the case of chromium sulphate: In dilute solution, this compound gradually dissociates first into its basic salts, and finally into the hydroxide, the breaking up of the neutral salt being induced by the presence of the fur-hair.

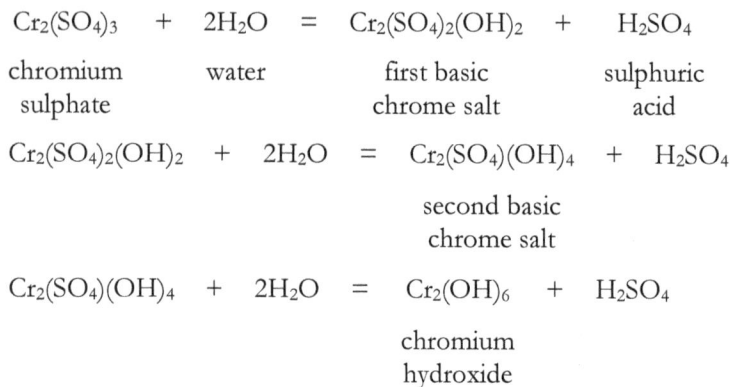

$$Cr_2(SO_4)_3 \quad + \quad 2H_2O \quad = \quad Cr_2(SO_4)_2(OH)_2 \quad + \quad H_2SO_4$$

| chromium sulphate | water | first basic chrome salt | sulphuric acid |

$$Cr_2(SO_4)_2(OH)_2 \quad + \quad 2H_2O \quad = \quad Cr_2(SO_4)(OH)_4 \quad + \quad H_2SO_4$$

second basic
chrome salt

$$Cr_2(SO_4)(OH)_4 \quad + \quad 2H_2O \quad = \quad Cr_2(OH)_6 \quad + \quad H_2SO_4$$

chromium
hydroxide

These reactions take place within the fibre, after the hair has been impregnated with the solution of the neutral salt, and when the compound has been rendered completely basic, in other words has reached the form of the hydroxide, it is supposed to combine with the acid groups contained in the hair substance, forming thus some complex, insoluble organic compound of the metal within the hair. According to some authorities the mordant is supposed to be present in the hair simply as the hydroxide, being tenaciously held by some physical means. The facts seem to indicate, however, that the metal is actually combined in some chemical way with the hair. For, if the mordant were present as hydroxide, then on white hair it would show the color of the hydroxide, which it does not. The same facts obtain with regard to other metals.

In order for the hair to be properly mordanted, it is necessary that the metallic compound which is taken up by the hair be held in such a manner that the mordant cannot be removed by water or even dilute acids or alkalies. Salts which dissociate too readily produce mordants which are only superficially precipitated on the hair and subsequently come off. Usually some substance is added to the solution of the salt to cause slower and more even dissociation of the salt, so that the hair substance can be quite saturated with the metallic compound before any insoluble precipitate is formed. Dilute sulphuric acid, organic acids like acetic and lactic, and cream of tartar are used to facilitate the uniform absorption of the mordant salt by the hair.

When the skins are mordanted before dyeing, they are immersed for 6 to 24 hours in a solution containing 1 to 20 grams of the metallic salts per liter of water, together with the corresponding quantity of the assistant chemical. The skins should be so entered into the mordant solution that the hair is uniformly in contact with the solution, and all the skins so that they are acted upon alike. Machinery such as is used for killing is suitable for mordanting also. The duration of the mordanting, and the concentration of the solutions are varied according to the depth of shade required, and also according to the nature of the dye to be employed. By suitably combining several mordants a considerable range of colors can be obtained with a single dye.

The various chemicals used as mordants are essentially the same no matter for which class of dyes they are used, there being only slight differences in the concentrations of the solutions, the manner of application of the mordants being practically the same. It is interesting to note that with the exception of chromium compounds, which are of comparatively recent adoption as mordants, all the chemicals now used for mordants were employed by the earliest masters of the art of fur dyeing. While some of the formulas used by those dyers display a lack of appreciation of the true action and function of the mordanting chemicals, yet it is quite remarkable that they chose, in spite of their limited knowledge of chemical processes and phenomena, just those materials which do act as mordants if properly applied. The most important metallic compounds for mordanting furs at the present time are salts of aluminum, iron (ferrous), copper, tin and chromium (as well as chromates and bichromates). The compounds of the metals with organic acids such as acetic acid are preferable, being more easily dissociated, and also leaving in solution an acid which is less injurious to the fur than a mineral acid. However, sulphates and other salts of the metals are also used extensively, inasmuch as they are cheaper than the organic salts.

ALUMINUM MORDANTS

Chief among the aluminum mordants are the various kinds of alum, which is a double sulphate of aluminum and an alkali such as sodium, potassium or ammonium. All these salts except that of sodium, form large, colorless, octahedral crystals, and are soluble in about 10 parts of cold water, and ¼ part of hot water. Sodium alum is even more easily soluble, but on account of the difficulty of obtaining it in crystalline form, it is little used. The common commercial alum is the potassium aluminum sulphate.

Recently, aluminum sulphate has to a large extent replaced alum for mordanting purposes, because it can be obtained very cheaply in pure form, and it contains a greater amount of active aluminum compound than does alum. Only the iron-free salt, however, may be used for the needs of fur dyeing.

Aluminum acetate also finds extensive application as a mordant in fur dyeing, and while somewhat more expensive than the alum or aluminum sulphate, it has the advantage over these compounds of being combined with an organic acid, which is preferable when the action on the hair and leather is considered. Aluminum acetate can be obtained in the market in the form of a solution of 10 degrees Beaumé, but can also be prepared very easily as follows:

665 grams pure aluminum sulphate, or

948 grams potassium alum, are dissolved in

  1 liter of hot water.

1137 grams of lead acetate (sugar of lead) are also dissolved in

  1 liter of hot water.

The two solutions are mixed, and thoroughly stirred. A heavy white precipitate forms, which is filtered off, and discarded after the solution has cooled. The aluminum acetate is contained in the filtrate, and the solution is brought to a density of 10 degrees Beaumé by the addition of water, if necessary, and is preserved for use in this form.

IRON MORDANTS

Ferrous sulphate, iron vitriol, or copperas, as it is commonly known, forms pale green crystals, which on exposure to air lose water, and crumble down to a white powder. It is very soluble in both cold and hot water, but the solutions oxidize very rapidly, turning yellowish, and should therefore be used immediately. Care must be taken that a good quality of iron vitriol be used for the mordant, otherwise very unsatisfactory results will be obtained.

Ferrous acetate is prepared in a manner similar to the aluminum acetate, and is occasionally employed instead of the ferrous sulphate. Inasmuch, however, as the solution of ferrous acetate is very easily oxidizable when exposed to the air, a more stable form is used, and this comes on the market as iron pyrolignite or iron liquor. This can be prepared by dissolving iron in crude acetic or pyroligneous acid, or by treating a solution of iron sulphate with calcium pyrolignite. Iron liquor is really a solution of ferrous acetate that contains certain organic impurities which prevent, or rather, considerably retard the oxidation of the iron salt, but which in no way interfere with its mordanting properties. The commercial product can be had in various concentrations, but 10 degrees Beaumé is the most usual and most convenient.

COPPER MORDANTS

The most important copper salts used in fur dyeing processes are copper sulphate, or blue vitriol, occurring in large blue crystals, very soluble in cold and in hot water; and copper acetate, which is formed by treating a solution of copper sulphate with a solution of the requisite quantity of lead acetate. Copper acetate can also be obtained in the form of blue-green crystals, very soluble in water, the solution becoming turbid on prolonged heating, due to the formation of a greenish basic copper acetate. This insoluble compound is known commonly as verdigris, although it is not usually produced in the manner mentioned. Numerous fur dyeing formulas contain verdigris, but inasmuch as the basic copper acetate is insoluble and thus incapable of reacting with any of the substances used in dyeing, it is assumed that the soluble normal copper acetate was meant, for this compound is also sometimes called verdigris.

In addition, there must be mentioned here a compound which formerly found extensive use in fur dyeing. This is a double salt of copper and iron, analogous to alum, ferrous copper sulphate, known as blue salt. It is very seldom used at the present time, being more effectively replaced by other substances.

CHROMIUM MORDANTS

The typical chromium mordant is chrome alum, which is a potassium or ammonium chromium sulphate, constituted just like the aluminum alums, and forming crystals like these. More frequently used, nevertheless, than the chrome alum, is chromium acetate, which is prepared from it, either by treating a solution of the chrome alum with a solution of lead acetate, or in the following manner:

50 grams of chrome alum are dissolved in

500 cubic centimeters of boiling water. To this is added

15 grams of 20% ammonia, diluted with 15 grams of water.

The precipitate which forms is filtered off, and preserved, the filtrate being discarded. After thoroughly washing the residue on the filter it is dissolved in dilute acetic acid, heating if necessary, to effect solution.

Other chromium compounds of an entirely different type are also used in fur dyeing, these being chromates and bichromates, the latter finding greater application than the former. Sodium bichromate is the salt most usually employed. This forms orange-red crystals which are very soluble in water, and in addition to its use as a mordant it also serves as an oxidizing agent for developing or fixing certain dyes on furs.

TIN MORDANTS

Compounds of tin find only limited application in fur-dyeing, the only one of importance being tin salts, stannous chloride, which occurs in the form of white, hygroscopic crystals, which must be preserved in closed vessels. It is very soluble, but in dilute solutions it readily forms a basic salt, so stannous chloride is usually used in very concentrated solutions.

ALKALINE MORDANTS

After the furs have been treated with the solution of some alkali for the purpose of killing the hair, they are always passed through a slightly acidulated bath to remove any alkali which may still be adhering. This operation must always be gone through before the skins can be mordanted or dyed, for if it were neglected, very uneven and uncertain results would be obtained. This process, however, entails the expenditure of no small amount of time, labor and chemicals when large lots of skins are being handled. In order to eliminate this extra step of "souring" between killing and mordanting or dyeing, it has been proposed to use alkaline mordants which combine the killing and mordanting functions, and accomplish these two processes at the same time. The advantages of employing such mordants are easily apparent. Cumbersome manipulation and handling of the skins, with the attendant consumption of much time and labor are reduced to a minimum, and besides there is no needless waste of chemicals as is the case in the ordinary methods of killing the furs.

The principle of alkaline mordants is not a strictly new one. If it be remembered that the old killing formulas used by the fur dyers of an earlier age, contained metallic salts with mordanting properties in addition to the alkaline substances, which alone were effective as killing agents, it would seem that the suggested alkaline mordants were merely a revival in modified form of the old processes. This is undoubtedly true in a large measure, for the killing mixtures which the old masters used certainly embodied the fundamental principle of simultaneous killing and mordanting, although it was not recognized at that time.

Modern alkaline mordants have therefore been devised which can be employed for killing and mordanting furs at the same time. They are prepared as follows:

ALKALINE ALUMINUM MORDANT

250 grams of potassium alum are dissolved in

  1 liter of boiling water. To this solution is added

300 grams of soda ash, previously dissolved in

750 c.c. of water, and the resulting precipitate is filtered off, washed and pressed, and then dissolved in a solution of 65 grams of caustic soda in 1 liter of water.

ALKALINE CHROMIUM MORDANT

250 c.c. of chrome acetate mordant of 20 degrees Beaumé

320 c.c. of caustic soda solution of 38 degrees Beaumé (32.5%)

  10 c.c. of glycerine 30 degrees Beaumé (95%)

The solution of these substances is brought up to a volume of 1 liter by the addition of 420 c.c. of water.

ALKALINE IRON MORDANT

138 grams ferrous sulphate are dissolved in

362 c.c. of warm water. Cool and add

  25 c.c. of glycerine. Then slowly and carefully add

25.5 c.c. of concentrated ammonia, taking care that no precipitate forms.

While these alkaline mordants seem to have much in their favor, there are certain possible objectionable features which must be considered. The solutions of the mordants are generally very alkaline, and not every fur can withstand more than a limited quantity of alkaline substance for longer than a comparatively short time. Suitable mordanting usually requires a longer time than killing does, so with the use of the alkaline mordant, if the skins remain in the solution until sufficiently killed, they may be insufficiently mordanted, while if the furs are treated long enough to be properly mordanted, the hair may have been over-killed. However, the idea of the alkaline mordant is a good one, and it is only a matter of time and patient, scientific experimentation when the difficulties of the method will be eliminated, and a much-desired process will become a practical realization.

The general methods for applying the various mordants of all sorts follow closely the procedure adopted for the killing formulas, and similar precautions must be observed, in order to obtain consistently uniform results. With the exercise of care, there is little reason for the mordanting operations to go wrong.

After proper treatment of the skins in the mordants, they are removed and drained off, then rinsed lightly in running water to remove the excess of mordant liquor, after which they can be directly entered into the dye bath. If it is not feasible to dye the mordanted skins at once, as is often the case, the skins are kept moist, and under no circumstances allowed to dry.

CHAPTER XII
FUR DYEING
MINERAL COLORS USED ON FURS

Before the introduction of the fur dyes now used, certain inorganic chemical substances were employed in addition to the vegetable dyes, for the production of colors on furs. Even to this day such materials are used to obtain certain effects in special instances. The idea of employing mineral chemicals undoubtedly originated in the textile-dyeing industry, which at one time was dependent to an appreciable extent on mineral substances for the production of certain fast shades. Compounds of iron, lead, manganese, also of copper, cobalt and nickel were all used for dyeing, either singly or in various combinations. In the application on furs, the brush method was the only one practicable, as the skins would have been ruined by dipping them into solutions of these chemicals in the concentrations necessary for dyeing.

The dyeing of furs with mineral colors involves the precipitation on the fibre in a more or less permanent form of the sulphide, oxide or other insoluble compound of a metal, and can be brought about in several ways. By what is known as double decomposition, that is, by the use of two solutions successively applied, the ingredient of one causing a precipitate to form when in contact with the constituent of the second, the color is produced on the hair. Another method is to use solutions of chemicals which decompose on contact with the hair, forming an insoluble compound. In the first method the hair is alternately treated with the two solutions of the requisite chemicals, drying between each brushing, the process being repeated until the desired shade is obtained. The second method merely requires the solution of the chemical to be applied to the hair, which is then dried, the color forming by itself.

One of the most important of the mineral dyes, and which is occasionally used to this day, is lead sulphide, formed by the double decomposition method by precipitating a soluble lead salt with ammonium sulphide, or any other alkaline sulphide. By simply brushing an aqueous solution of lead acetate, also known as sugar of lead, on a white fur such as white hare or rabbit, a light, brownish coloration is obtained due to the combination of the lead with the sulphur of the hair. If the lead solution is carefully applied several times on this type of fur, until a sufficiently dark color is produced, it is possible to get a fairly good imitation of the stone marten. The brown color is very fast, being actually formed within the hair. In most cases, however, for dyeing lead sulphide shades it is necessary to use the two solutions. Thus the pale greyish or slightly brownish-grey shades of the lynx can be reproduced on white rabbit or hare by this process. A solution containing 60 grams of lead acetate per liter of water is brushed on to the hair of the fur which has previously been killed in the usual manner, and the hair is then dried. A solution of 50 grams of ammonium sulphide per liter of water is next brushed on, and the fur again dried. Care must be exercised in handling the ammonium sulphide as it is a very malodorous liquid, the fumes of which are poisonous when inhaled. The alternate brushings are repeated until the desired depth of shade is obtained. A very dark brown, approaching a black can be obtained in this way. This color can be used for the production of certain attractive effects. By brushing over the tips of the hair, which has previously been dyed a dark brown by means of the lead sulphide color, with a dilute solution of hydrochloric acid, or with peroxide of hydrogen, the hair will become white in the parts so treated, due to the formation of lead chloride or lead sulphate, respectively. Thus white tipped furs can be obtained, but the process is applicable only when the furs have been dyed by the lead sulphide method.

Potassium permanganate is occasionally used to produce dyeings of a brown shade on furs. Considerable care has to be taken in applying this substance, as it is possible to affect the hair. The strength of the solution must be varied according as the hair to be dyed is weak or strong. A cold solution of 10 to 20 grams of potassium permanganate per liter of water is brushed on to the hair, which is then dried. A brown precipitate of manganese is formed on the hair after a short time, and the process is repeated until the required shade is obtained. For furs with harder hair, stronger solutions can be used. The dyeing is very fast, but it is seldom used, cheaper and better shades being obtained in other ways. Spotted white effects can be produced on the brown dyeing with permanganate of potash by applying a solution of sodium bisulphite, the brown color being dissolved by this chemical.

The compounds of other metals, such as iron, copper, cobalt and nickel are not used in practise as the dyeings are not fast, and can be better produced in other ways.

CHAPTER XIII
FUR DYEING
VEGETABLE DYES

With the exception of the few shades which could be produced solely by means of coloring matters of a chemical character, all dyeings on furs up to about thirty years ago were made with dye substances obtained from the vegetable kingdom, either alone, or in conjunction with the aforementioned mineral colors. The colors of vegetable origin used in comparatively recent times were mainly extracts of the wood of certain trees; so the name "wood dyes" has come to be applied generally to the dyes of this class. The use of the vegetable or natural dyes on furs dates back to quite ancient times, as frequent allusions and descriptions in Biblical and other contemporaneous literature testify. There are numerous pictures on monuments and tablets illustrating the dyeing of furs among the ancient Egyptians, the evidence indicating that the juice of certain berries, and extracts of certain leaves were used for the purpose. At a later period, in the Roman era, henna, which was used over two thousand years ago as to-day for the beautification of the hair of women, was also used to color fur skins. The instances cited here are merely of scientific and historical interest, and are not of practical importance as far as fur dyeing methods are concerned.

It was not until many centuries later that the dyeing of furs took on the aspects of a commercial art, and the substances then employed were chiefly tannin-containing materials such as gall-nuts and sumach, which in conjunction with certain metallic salts, particularly those of iron, were capable of producing dark shades. The use of iron compounds to form dark grey or black colors on leather tanned by means of the tannins, had been common for a long time, and it was natural that fur dyers should try to produce such shades on furs in a similar fashion. The use of the iron-tannin compound as a dye proved to be very effective, and to this day the production of blacks by means of the vegetable coloring matters has as a basis an iron-tannate. A formula in common use in the latter seventeenth and the eighteenth centuries for producing black shades on furs, is the following:

Lime water	1117	parts
Gall-nuts	1500	,,
Litharge	500	,,

Salammoniac	65	,,
Alum	128	,,
Verdigris	64	,,
Antimony	64	,,
Minium	32	,,
Iron filings	128	,,
Green copperas	384	,,

All these substances except the gall-nuts, the copperas and half the lime water were boiled up in a cauldron; then the gall-nuts and the copperas were placed in a bucket and the contents of the cauldron poured in, and the rest of the lime water added. The mixture was stirred up, allowed to settle for an hour, and when cool, was ready to be applied by the brush method. For dyeing by the dip process, a similar mixture was used, only considerably diluted with water. A study of the formula discloses the fact that in it are combined killing and mordanting substances as well as dyeing materials. The lime water, in conjunction with the salammoniac serves as a killing agent, the verdigris, copperas and alum are mordants, while the litharge and the minium, both compounds of lead, could possibly act as mineral dyes, and the iron filings and the antimony took virtually no part at all in the dyeing, except, perhaps to act in a mechanical way.

The formulas for other shades were made up along similar lines, the chief constituent of vegetable nature being either gall-nuts, sumach, or both. A mixture for a chestnut brown, for example, contained gall-nuts, sumach, and the various other mineral constituents as in the black dye, litharge, alum, copperas, verdigris, salammoniac, antimony, and in addition, red lead and white lead. It is evident in both these instances that the shade obtained was as much the result of mineral dyeing as of vegetable dyeing.

The discovery of America introduced into Europe many new dye substances, chiefly wood extracts such as logwood and Brazilwood, but it was not until the nineteenth century that these materials found their way into the dye formulas of the fur dyer. Most of the processes used in the dyeing of furs were adaptations of methods employed in silk dyeing, the silk fibre being considered as most nearly approaching fur-hair in nature and characteristics. By devious and circuitous paths the formulas of the silk dyers reached the fur people, and so, in the middle of the nineteenth century, dye mixtures containing the various dyewoods as well as the tannin-containing substances were in general use for the dyeing of furs. The following is a typical recipe of that time for the production of black on furs like wolf, skunk, raccoon, etc.:

Roasted gall-nuts	1000	parts
Sumach	200	,,
Iron mordant	200	,,
Copper vitriol	100	,,
Litharge	80	,,
Alum	60	,,
Salammoniac	50	,,
Crystallized verdigris	40	,,
French logwood extract	30	,,
Rain water	7000	,,

The mixture was boiled up, and after cooling was ready for application by the brush method, the skins being first killed by a killing mixture also applied by the brush. The dye substances in this case are the gall-nuts, sumach and the logwood extract, with the iron mordant, copper vitriol, and alum as mordants. For brown shades a similar formula was used containing Pernambuco wood extract, logwood extract, quercitron bark, gall-nuts and dragonblood, together with iron, copper and alum mordants.

Formulas such as the above were mainly empirical, that is, they were compounded as a result of trial of various combinations of the constituents, without considering the nature and quantitative character of the reactions, as long as the desired shades could be obtained. Such dye mixtures were frequently found to yield results varying from those expected or originally obtained, because the effectiveness of the formulas depended upon the exact duplication in every detail, of conditions which had given satisfactory results previously, and it was not always possible to attain such an accurate reproduction of circumstances, especially when the fur dyers were quite ignorant of the scientific relationships of the materials used. So when more light had been shed on the nature and chemical characteristics of the vegetable dye substances, formulas like those described were no longer employed, although the essential ingredients were the same in the new processes. Unnecessary constituents were eliminated, and proper ones substituted where it was required, and the quantities of the materials used were made to conform to the chemical laws governing the reactions. Since these new formulas were based on a rational understanding of the constituents and their reactions, it is desirable to study the latter briefly, before further discussing the formulas themselves.

The substances of vegetable origin used in modern fur dyeing may be grouped into two classes, one, the tannin-containing materials, and the other, the dyewoods proper. The most important of the tannins are gall-nuts, sumach and chestnut extract. Cutch, which also comes under this class, is more frequently used for the production of brown shades, so it is grouped with the dyewoods. Among the latter are logwood, fustic, Brazilwood, quercitron, turmeric, and several others of less significance.

1. *Tannin Materials*

First and foremost under this heading are the nutgalls. These are ball-shaped excrescences produced on certain plants by the punctures of insects in depositing their eggs. There are two chief varieties, the European, and the Chinese. The European galls are formed by the female gall-wasp which drops an egg in the rind of young branches of certain oaks. A swelling (the nutgall) is produced, in which the young insect develops, and from which it finally escapes by piercing a hole through the shell. Those galls which are not pierced have a fresh bluish or green color, are heavy and contain most tannic acid. After the insect has gone out, the galls are of a lighter, yellowish color, and also of inferior quality. The best oak-galls are the Aleppo, and the Turkish or Levant galls, containing 55–60% of tannic acid, and about 4% of gallic acid. The Chinese galls are produced by the puncture of a plant-louse on the leaves and leaf-stalks of a species of sumach, and not on oaks. The galls are very light, and very rich in tannic acid, containing often as much as 80%. For dyeing purposes, nutgalls are usually ground to a powder, and in some instances they are even roasted first and then ground.

Sumach consists of the leaves and sometimes of the small twigs and stems of a species of sumach plant known as the Rhus coriaria. The Sicilian variety is the finest commercial quality, with the Virginian ranking next. It is sold as a powder, but also in the form of the whole or crushed leaves. The best sumach contains 15–25% of tannin. Extracts are also manufactured, a liquid extract of 52 degrees Twaddell, which forms a dark brown, thick paste; and a solid extract, formed by evaporating the liquid extract to dryness.

Chestnut extract is prepared from the wood of the chestnut oak, which contains 8–10% of tannin. The solid extract has a bright, black color, while the liquid extract is a dark brown paste with a smell like that of burnt sugar.

The tannins all give greyish to black shades with iron salts, and it is this fact which renders them important for fur dyeing.

2. Wood dyes

One of the most important of all the natural dye substances, especially for the production of blacks, is logwood. The color is really a red, but with the common mordants it forms blue, violet or black shades. Logwood, or campeachy wood, as it is sometimes called, is the product of a large tree growing in the West Indies, and Central and South America. When freshly cut, the wood is practically without color, but when exposed to the air it soon becomes a dark reddish-brown on the surface. The coloring principle of logwood is called hematoxylin, which is a colorless substance when pure, and is of itself incapable of dyeing; but when it is exposed to the air, especially when moist and in the presence of some alkaline substance, it is converted into hematein, which is the real coloring matter of logwood. To prepare the wood for use, the logs are chipped or rasped, the chips being heaped up and moistened with water. Fermentation occurs, and the heaps are frequently turned to allow free access of air to the wood, and to prevent overheating. As a result of this process, a great part of the hematoxylin is converted to the hematein. The logwood may be used for dyeing in this state as chips, but logwood extracts can now be obtained of a high degree of purity and are easier to work with. The commercial forms of the extract, are the liquid of 51 degrees Twaddell, and the solid extract. Hematein crystals can also be obtained. All these extracts contain mainly hematein, together with a small percentage of hematoxylin which is converted to the former during the dyeing process. Logwood is never used as a direct dye, but is used to form color lakes with the various mordants, the following colors being produced:

- Iron mordants give grey to black shades

- Copper mordants give green-blue to black shades

- Chrome mordants give blue to black shades

- Aluminum mordants give violet shades

- Tin mordants give purple shades

By combining several of the mordants, any desired shade of black can be obtained, and if other dyewoods are used in conjunction with the logwood, the range can be further increased.

Fustic, yellow-wood, or Cuba wood, as it is variously called, is obtained from a tree also growing in the West Indies, Central and South America. It is used either as wood chips, or as a paste extract of 51 degrees Twaddell, and occasionally as solid extract. Fustic contains two coloring matters, morintannic acid, possessing the characteristics of a tannin, and which is quite soluble in water, and morin, which is rather insoluble, and which settles out from the liquid extract. Fustic is the most important of the yellow dyes of natural origin, and is used considerably in fur dyeing with logwood for shading the blacks, or for producing compound shades. With the usual mordants fustic gives the following colors:

With iron salts	dark olive
With copper salts	olive
With chrome salts	olive-yellow to brownish-yellow
With aluminum salts	yellow
With tin salts	bright yellow to orange-yellow

Brazilwood, or redwood, is the product of a tree found in Brazil, and exists in several varieties, such as peach wood, Sapan wood, Lima wood, and Pernambuco wood. They all yield similar shades with the various mordants, and all seem to contain the same coloring principle, brasilin, which, like the hematoxylin, has no dyeing power, but by fermentation and oxidation it is converted to brasilein, corresponding to the formation of hematein. Brazilwood and the related woods are used either as chips or extract, but seldom alone, usually in conjunction with other dyewoods. By combining logwood, fustic and Brazilwood in various proportions, and by employing suitable mordants, all the shades required by the fur dyer can easily be produced.

Quercitron is the inner bark of a species of oak (Quercus tinctoria) found in the United States. It contains two coloring principles, quercetrin and quercetin. The fresh decoction of quercitron bark is a transparent dull orange-red which soon becomes turbid and deposits a yellow crystalline mass. It is generally used in conjunction with other dyes.

Cutch is the dried extract obtained from a species of acacia, the principal varieties being Bombay, Bengal, and Gambier cutch. It contains two coloring principles, catechin and catechu-tannic acid. Cutch acts as a tannin, and like other tannins discussed above, can be used for the production of grey or black shades with iron mordants. It is employed chiefly, however, for dyeing browns. Aluminum salts give with cutch a yellowish-brown, tin salts give a lighter yellow, copperas gives a brownish-grey, and chrome and copper salts give brown shades.

Turmeric is the underground stem of the Curcuma tinctoria, the coloring principle being called curcumin. It may be used as a direct dye, but usually a mordant is used. Turmeric is sometimes used in place of fustic.

While the tannins can be used alone with an iron mordant for producing greyish to black shades, the dyewoods alone yield colors which would be too bright to be suitable for dyeing furs. In order to tone down this brightness, and to give to the dyeings that greyish undertone which is characteristic of the natural furs, and which can only be imitated by means of the iron-tannin compound, it is customary to combine the tannins with the wood dyes. The iron-tannate constitutes the foundation of the color which gets its intensity, and necessary brilliancy and bloom from the wood dyes. Moreover, the presence of the iron-tannin compound helps considerably to increase the fastness of the dyeing. Furs dyed with the combination of the tannins and the wood dyes obtain an additional tanning treatment which materially improves the quality of the leather, for not only do the tannin substances exert this tanning action, but the dyewoods as well, for they are themselves either of the nature of tannins, or contain a coloring principle which is a tannin. It is to the combined effects of the tannin substances and the dyewoods that furs dyed with vegetable dyes owe their beauty of color, lustre, naturalness of shade, permanence of the dyeing, and durability of the leather. Wood dyeings on furs have for this reason acquired a just renown, but owing to the introduction of the new kinds of fur dyes, the use of the vegetable dye substances has been greatly reduced.

The dyes of vegetable origin can be applied to furs by either the brush method or the dip method, or both, and since mordants are required with the dyes of this class, they are applied in one of the three ways mentioned in a previous chapter: first, by mordanting before dyeing; second, by applying mordant and dye simultaneously; and third, by mordanting after the skins have been treated with the dye.

I. DYEING WITH VEGETABLE DYES BY THE BRUSH METHOD

The use of the brush method in applying the natural dyes to furs is limited to a comparatively few kinds of dyeing, namely to produce special effects on furs, or to give to the upper-hair of furs a coat of dye different from the base color. In a quite recent German patent is described a process for blending a red fox as a silver fox and the procedure affords a good example of brush dyeing with preliminary mordanting. The specification is as follows: "D. R. P. 310, 425 (1918). A process for dyeing red fox as silver fox. The tanned and dressed skin is first superficially decolorized by applying a dilute mixture of milk of lime, iron vitriol and alum, with a soft brush so as only to penetrate the top-hair. Allow to remain for 4–6 hours, dry, and beat out the dust. A dilute solution of iron vitriol is brushed on so as only to wet the top-hair, and the skin is thus allowed to remain moist for 12–24 hours. Then without drying, a solution of iron vitriol, salammoniac, litharge, red argol and wood ashes is brushed on cold with a hard brush so as to penetrate all the hair down to very near the skin. The skin has now completely lost its red color, and has become a pale yellow. It is now ready to be dyed. An infusion of roasted nutgalls, which have been boiled for 3–4 hours with water, is applied cold with a soft brush to the upper hair. Allow to remain so for 2–3 hours, and without drying, apply a weaker solution of the roasted nutgalls with a hard brush so as to saturate the hair thoroughly. Dry and beat out. According to the concentration of the solution applied, the hair will be colored blue-grey to black, and the shade can be varied by varying the strength of the solutions used. The different parts of the skin, or those parts of different shades can be dyed accordingly."

In this patent all the operations, including killing, mordanting and dyeing are done by the brush method, and the process, from this point of view is quite similar to one which might have been employed a century previous. It is evident that the time and effort required to carry out the details as described in the patent would only be warranted in exceptional cases, where the value of the dyed fur would be considerably greater than that of the natural skin.

An example of the application at the same time of dye and mordant by the brush method is the original French Seal dye, which is still employed to a limited extent to produce a brilliant, deep, lustrous black topping on furs which have already been dyed by the dip process. A typical formula for the old French Seal dye is the following:

Green copperas	10 parts
Alum	10 ,,
Verdigris	10 ,,
Gall-nuts	80 ,,
Logwood extract (15 degrees Twaddell)	150 ,,
Water	1000 ,,

This mixture is applied to the top of the hair of the furs, after previous killing, and the skins allowed to remain moist for several hours, and also exposed to the air. The skins are then dried, and beaten out, and if necessary a second coat of dye is brushed on. In dyeing seal-imitation on muskrat, or skunk-imitation on opossum, for example, the black color required on the top-hair, or the upper part of the hair when the furs are sheared, can be produced by applying a mixture similar to the above, to the furs after they have received their base color by the dip process with natural dyes or with the Oxidation Colors. Occasionally, the dyeing is given an after-treatment with a dilute solution of sodium bichromate to help develop the color, the action in this case being that of an oxidizing agent, and not of a mordant.

As far as the third method of mordanting is concerned, that of first applying the dye, and then the mordant, it is rarely practised with the brush method. The procedure, however, consists in first brushing on a solution of the desired dye, then drying and brushing on a mordant solution. These operations are repeated perhaps two or three times until the proper shade is obtained, exposing the furs to the air for the color to be developed.

II. Dyeing with Vegetable Dyes by the Dip Method

It was in the application to furs by the dip process that the use of the vegetable dyes attained great importance, and although at the present time, natural organic dyes have largely been superseded by the Oxidation Colors and Aniline Black dyes, yet for certain purposes, and especially for the production of blacks, the wood dyes still are able to hold their own.

The dyeing of black formerly constituted probably the most important branch of the fur dyeing industry, and was undoubtedly the most difficult one. For it is possible to obtain as many different kinds of black as there are dyers of this color, but only a few certain shades are desirable. The division of the classes of furs into those derived from the various kinds of sheep, and those obtained from other animals is particularly marked in the dyeing of black, and both the composition of the dye formulas and the methods of dyeing are somewhat different for the two groups. For the dyeing of black on Persian lambs, broadtails, caraculs, etc., a combination of logwood and nutgalls with the requisite mordants is used, while on hares, Chinese sheep, foxes, raccoons, opossum, etc., a mixture of logwood and turmeric or fustic, with the proper mordants is used.

The general procedure is as follows: The dye substances to be used are ground up to a powder in a mill constructed for the purpose, after which they are boiled with water in a copper-lined kettle or cauldron, heated from the outside by steam. The customary arrangement is to have a jacketed kettle, supported on a stand, and having taps and valves to enable the liquor to be drawn off, or pivoted, so that the kettle can be tilted, and the contents poured out. The use of the copper-lined vessel is to be preferred, as it is unaffected by any of the dye substances, and so cannot cause any rust stains. After the dyes have gone into solution and have cooled, the mordant chemicals, previously dissolved in water, are added, and the mixture stirred up. The dyeing in this instance is effected by the simultaneous application of dye and mordant. The dye mixture is now run off, or poured out in the proper quantity into a number of small vats of 25–30 gallon capacity, or into a paddle vat, which can be closed, while the paddle is rotating. The latter device is to be preferred because it permits the dye to retain its temperature better and for a longer period of time, but when lambs are being dyed only the open vats are used. The temperature of the dye mixture is between 40° and 45° C., for only at this temperature can the hair absorb the dye properly without injuring the leather. The killed skins are immersed in the dyebath for a time, usually overnight, after which they are removed, drained and hung up, with the hair-side exposed to the air, so as to permit the dye to develop, which takes place with the aid of the atmospheric oxygen. The dyebath is again brought to the proper temperature, and the skins are again entered, to go through the same process as often as is necessary to obtain the desired depth of shade. The dyed skins are thoroughly washed to remove excess dye, then dried and finished. The following are a few dye formulas used in the production of blacks:

Logwood extract   100 grams

Chestnut extract   14 c.c.

Turmeric   38 grams

Iron acetate 6° Bé   50 c.c.

Water   1200 c.c.

or,

Cutch   15 grams

Soda   14 grams

Logwood extract   120 grams

Verdigris   19 grams

Iron acetate 5° Bé.   16 c.c.

Water   1200 c.c.

A recently published formula for dyeing China goat skins black, is the following:

Dissolve 50 lbs. of dark turmeric and 45 lbs. of logwood extract and make up to 300 gallons of solution, at 95° F. Enter the killed skins and leave them in the liquor until they rise to the surface. Then take them out and add 25 lbs. of logwood extract, 10 lbs. of sumach, 10 lbs. of blue vitriol, 5 lbs. of fustic extract, and about 60 lbs. of iron acetate liquor. Stir up well, and immerse the skins for 18 hours. Draw them up, and expose to the air for 12 hours. Heat the liquor again to 95° F. and put the skins back for 12 hours.

Draw out, hang up in the air for a time, then wash thoroughly, hydro-extract, dry and finish.

In a German patent, D. R. P. 107,717 (1898), is described a method for dyeing lambs black, consisting in treating the skins for 24 hours in a logwood bath, then rinsing in cold water, and mordanting for 15 hours in a solution of bichromate of potash. The skins are then washed and treated with a solution of iron salt, then dried. This process, while of not much practical importance, is an illustration of mordanting subsequent to the dyeing treatment.

As far as the production of other shades is concerned, the procedure is quite similar to the regular black method. For a dark brown, for example, the skins are dyed in a mixture containing

Gall-nuts	40 parts
Verdigris	10 ,,
Alum	10 ,,
Copperas	5 ,,
Brazilwood extract (15° Twaddell)	150 ,,
Water	1000 ,,

employing operations just as in the case of the black.

Greyish-blue shades on white hares, lambs, kids, etc., can be obtained by treating the skins successively in the following baths:

1. Logwood extract 100 grams

Water	1 liter

2. Indigotine 10 grams

Alum	10 grams
Water	1 liter

Bluish-grey tones on the same furs can be produced by treating with

1. Logwood extract 200 grams

Indigotine	15 grams
Water	1 liter

2. Alum	150 grams
Salammoniac	12 grams
Water	1 liter

Similar grey shades can be produced by mordanting the skins with an iron salt, and then dyeing in a weak bath containing gall-nuts, sumach and iron vitriol. This method is very effective for making Alaska or silver fox imitations.

CHAPTER XIV
FUR DYEING
ANILINE BLACK

Fur seal for a long time has been a fur of distinction and importance in the fur industry, and consequently the dyeing of seal has constituted an important, though not very extensive branch of the art of fur dyeing. In quite recent times the popularity of seal has become so great that imitations have had to be produced to help supply the demand, and as a result, French seal, or seal-dyed rabbit, and the so-called Hudson seal, which is seal-dyed muskrat, have acquired a great vogue. Occasionally opossum, nutria and other furs are also used for the purpose of producing seal imitations. While the supply of real seals is relatively small, and the demand large, the production of seal imitations has assumed large proportions, and as a result, the dyeing of seal and its imitations or substitutes has come to be a great branch of the fur dyeing industry.

During the past thirty years, the long and tedious processes of dyeing seal and seal imitations, involving the use of dyes of vegetable origin, have largely been superseded by what is known as the Aniline Black dye. It was the French who first worked out successfully the application of Aniline Black to furs, and the method has attained much importance and extensive use in the fur dyeing industry.

Aniline Black is the name given to an insoluble black dyestuff produced by the oxidation of aniline in an acid medium. As a finished product it cannot be used in fur dyeing, but if the hair of the furs be impregnated with a suitable preparation of aniline and then treated with certain oxidizing agents, the color will be formed on the hair, being firmly fixed and giving a fast black, resistant to light, washing and rubbing. The basis of the dye, aniline, is an oily liquid, possessing a peculiar fishy odor, colorless when pure, but rapidly turning brown when exposed to the air. It is obtained from benzol, which is distilled from coal-tar, by treating with nitric acid, forming nitrobenzol, which when subjected to the action of reducing chemicals is converted into aniline. The process may be shown schematically as follows:

Coal—coal-tar—benzol—nitrobenzol—aniline oil—Aniline Black. Aniline Black was by no means a new dye when the French succeeded in producing it on furs. It had been used for a long time previous on textiles, chiefly cotton. The history of the development of the Aniline Black process throws considerable light on its nature and constitution, and so presents many features of interest. As early as 1834, the chemist Runge observed the formation of a dark green color when heated aniline nitrate in the presence of cupric chloride. Fritsche, in 1840, noticed that when chromic acid was added to solutions of aniline salt, a dark green, and sometimes a blue-black precipitate was produced, and later the same chemist obtained a deep blue by the action of potassium chlorate on aniline salt. It is interesting to note that Perkin, in 1856, conducting similar experiments on the oxidation of aniline with chromic acid, obtained a blue-black product from which he extracted the first synthetic coal-tar dye, mauve. Thus far, all the experiments on the oxidation of aniline proved to be merely of scientific interest, but in 1862, Lightfoot patented a process for the practical application of colors formed by the oxidation of aniline on the fibre, a greenish shade being obtained by that method, to which the name emeraldine was given, and by subsequent treatment with bichromate of potash, the green was changed to a deep blue color. Since that time, the methods for producing and applying Aniline Black have been developed and improved, although all the processes were based on the principles incorporated in Lightfoot's original patent. However, it was not until the last decade of the nineteenth century that the dyeing of furs by means of the Aniline Black method was successfully attempted.

A knowledge of the nature and the manner of the chemical changes which take place in the production of Aniline Black is a valuable aid in obtaining satisfactory results in practise; and although Aniline Black was extensively used before the true character of the reaction was understood, since the successful determination of the constitution of Aniline Black and the discovery of the real nature of the process by Green and his collaborators in 1913, the methods have been considerably improved and simplified, with correspondingly better results in dyeing. As a consequence, the methods of dyeing furs with Aniline Black have also become simpler and more efficient.

A discussion of the chemical changes which occur in the Aniline Black process, is out of place here on account of the highly involved and complicated character of the reactions, to understand which requires a considerable knowledge of specialized organic chemistry. But the essential features of practical importance in the production of Aniline Black are the following: As already noted, one of the characteristic properties of aniline is its tendency to turn from a colorless to a dark-brown liquid in the presence of the air. This change is due, together with certain other causes, to an oxidation brought about by atmospheric oxygen. By employing oxidizing agents, this oxidation can be accelerated and carried further, and eventually the Aniline Black is obtained. Among the substances which may be used to bring about the conversion of aniline to the insoluble black dye are manganese dioxide, lead peroxide, hydrogen peroxide, chromic acid, ferric salts, potassium permanganate, chloric acid and chlorates in the presence of certain metallic salts, particularly those of vanadium and copper. Chlorates, especially sodium chlorate and potassium chlorate, are the most commonly employed oxidizing agents, bichromate of soda or of potash being used, in addition, to complete the oxidation. When using chlorates it is necessary to have present in the dye mixture a small quantity of a metallic salt, which, while not entering into the reaction itself, is nevertheless indispensable as an oxygen carrier. Vanadium compounds have proved to be the most effective for this purpose, and according to an authority, one part of vanadium salt is sufficient to cause the conversion of 270,000 parts of aniline to Aniline Black, the necessary amount of a chlorate being present of course. Salts of copper, cerium, and iron are also extensively used, but they are not quite so efficient as vanadium.

The formation of the Aniline Black in practise takes place in three well-defined steps, which it is important to be able to recognize and distinguish in order to obtain the best results. The first stage of the oxidizing process produces what is called emeraldine, which in the acid medium of the aniline bath is of a dark green, while in the free state it is of a blue color. As the oxidation proceeds, the second stage develops, the emeraldine being converted to a compound called nigraniline. This in acid solution is blue, and the free base is a dark-blue, almost black. It was formerly considered that the nigraniline was the Aniline Black proper, and so when this stage of the oxidation was reached, the process was often interrupted and not carried to the limit. This can account for the fact that Aniline Black dyeings usually turned green after a short time. The reason for this is that nigraniline, when treated with weak reducing agents, as, for example, sulphurous acid, is at once changed to emeraldine, with its dark green color. Since there is usually a small amount of sulphurous acid in the air, especially in places where coal or gas is burned, an Aniline Black dyeing which has not been carried beyond the nigraniline stage will be reduced in time to the emeraldine, and cause the dyeing to become green. The last step in the oxidation changes the nigraniline into what is properly called the ungreenable Aniline Black. Weak reducing substances like sulphurous acid do not change this compound to emeraldine, and stronger reducing agents only convert it to a brownish compound, which changes back to the black when exposed to the air. It is quite evident that in order to obtain a black which will not change to green in time, the oxidation of the aniline must be carried to the last stage. By making tests during the dyeing of the furs, it can easily be determined whether the oxidation has proceeded far enough.

In the dyeing of textiles with Aniline Black, it is customary to carry out the operation at comparatively high temperatures, approaching 100° centigrade. With furs such temperatures are out of the question, so it is necessary to repeat the dyeing several times in order to obtain the proper depth of shade working in the cold. Only the brush method can be used in applying the Aniline Black dye to furs, on account of the strong acidity of the dye mixture, which would ruin the leather, if the dyeing were done in a bath. Indeed, great care must be exercised even by the brush method to avoid too great penetration of the dye liquid, otherwise the roots of the hair will be attacked, and the leather may be "burned" from the hair side. Furs dyed with Aniline Black are frequently after-dyed by the dip-process with logwood or some other similar dye, in order to add to the brilliancy of the dyeing. Combined with intensity of color, Aniline Black on furs is the only dye which will also give fast, lustrous shades, and leave the hair soft and smooth.

There are several methods of applying Aniline Black on furs, the most important being

- 1. One-bath Aniline Black

- 2. Oxidation Aniline Black

- 3. Diphenyl Black

- 4. Aniline Black by Green's Process

1. One-bath Aniline Black

A typical formula for this method is the following given by Beltzer:

Aniline salt   10 kg.

Sodium chlorate   1.5 kg.

Copper sulphate   0.7 kg.

Vanadate of ammonia 10 gr.

All these substances are dissolved hot in 50 liters of water, and allowed to cool, forming solution A. Aniline salt is aniline oil which has been neutralized with the exact quantity of hydrochloric acid to form the hydrochloride. It forms white or greyish crystalline lumps very easily soluble in water. The sodium chlorate is the oxidizing agent, and the copper sulphate and the vanadate of ammonia are the oxygen carriers.

15 kg. of sodium bichromate are also dissolved in 50 liters of water, forming solution B. The bichromate is also an oxidizing agent and serves to complete the oxidation of the aniline to the black.

Immediately before using, solutions A and B are mixed together, both being cool. In general practise it is customary to mix only small quantities at a time, as a considerable precipitate forms when the whole batch is mixed at once, the precipitate being so much waste dye substance. Usually a liter of A and a liter of B are mixed at a time, and the furs brushed with the mixture. The brushing must be varied according as the hair is hard and stiff, or soft and tender. The hair must be thoroughly impregnated in all directions, and the penetration must not be too deep to affect the leather. With experience and dexterity satisfactory results can easily be achieved. After the skins have been properly treated, they are dried at a temperature of about 35 degrees centigrade. When dry, they are returned to the dye bench, where they receive another application of the dye mixture, and are again dried. This operation may be repeated as often as six or seven times before a sufficiently intense black is obtained. Another way of producing the desired depth of shade with fewer applications is by using more concentrated dye mixtures. Each method has its disadvantages, the greater number of brushings requiring the expenditure of more time and labor, and the greater concentration of the bath resulting in a considerable loss of dye substance due to the formation of a large precipitate when the two solutions are mixed, and moreover, not all furs can be treated with concentrated mixtures. The best results with this method usually require the application of six coats of a mixture of moderate concentration.

2. *Oxidation Aniline Black*

In order to overcome the difficulty of employing very concentrated dye mixtures, or of making many applications of the dye, a method was devised whereby the two solutions of the previous process, instead of being mixed together, are applied successively to the hair of the furs, the following formula, also by Beltzer, being an example:

Aniline oil   10 liters

Nitric acid 36° Beaumé, or

Hydrochloric acid 22° Beaumé 20 liters

Cold water   20 liters

This is solution A, and is merely a solution of aniline hydrochloride, or nitrate, depending on which acid has been used. Nitric acid, although more costly than the hydrochloric acid, is to be preferred, because it is an oxidizing acid, and so assists in the oxidation of the aniline, and besides, has a more beneficial effect on the hair than the hydrochloric, in the matter of softness and luster.

Sodium chlorate   4 kg.

Copper sulphate   1 kg.

Vanadate of ammonia 10 gr.

Water   50 liters

This is solution B, containing the oxidizing agent, and the oxygen carriers. Just before using, equal quantities of A and B are mixed, and the skins brushed with the mixture. The skins are then dried at 35–45° centigrade, at which temperature the color begins to develop. When almost, but not entirely dried, the skins are subjected to the action of warm vapor, which is allowed to enter the drying chamber, so as to keep the temperature about 40° centigrade, the color developing better in this way. This operation may be repeated, or the skins are directly treated with a solution of 25 kg. of sodium bichromate in 100 liters of water, to complete the oxidation. The moist skins are exposed to the air for a time, and then dried at 35° C.

This method of dyeing has several advantages over the One-bath Aniline Black. It requires fewer brushings, and enables the complete utilization of the dye solutions without loss. With three applications of the dye mixture by the Oxidation process, as deep and intense a black can be obtained as with six brushings by the One-bath method. The dyeings, too, are nearly, but not fully as brilliant and even as in the latter case. The greater the number of coats of dye that are applied the more regular will the dyeing be.

3. *Diphenyl Black*

In 1902, the Farbwerke Hoechst, a large German producer of coal tar intermediates and dyes, invented an Aniline Black process to which they gave the name Diphenyl Black. The chief departure from the previous Aniline Black methods was the replacing of part of the aniline oil of the dye mixture by Diphenyl Black Base I, which is para-aminodiphenylamine. This base has the property of being oxidized to Aniline Black, just like aniline oil, and the advantage claimed for the Diphenyl Black is that it produces an absolutely ungreenable black. The method of application is practically the same as for the other Aniline Black processes, chlorates being used as the oxidizing agents, in the presence of oxygen carriers such as salts of copper and vanadium. The use of bichromates is dispensed with. On account of the comparatively high cost of the Diphenyl Black Base I, this method has not found very extensive application, especially as highly satisfactory ungreenable blacks can now be produced by other methods.

4. *Aniline Black by Green's Process*

In 1907, Green, who has done much work in the direction of elucidating the character of the Aniline Black process, obtained a patent for a method of applying Aniline Black in a manner which was different from all the previously known formulas. The invention created great interest, and although in its original form it did not find a wide application, many of the methods used at the present time are in one way or another derived from the idea of Green. A resumé of the patent will therefore be given here: "The invention relates to the production of an Aniline Black, the new process differing from all other known processes by the fact that the oxidation of aniline is effected solely or mainly by the oxygen of air. The possibility of dispensing with an oxidizing agent depends on the discovery that the addition of a small quantity of a para-diamine, or of a para-amido-phenol to a mixture containing aniline and a suitable oxygen carrier, such as a salt of copper, greatly accelerates the oxidation of the aniline by the atmospheric oxygen. Further, whereas in the ordinary processes of Aniline Black, the quantity of mineral acid employed cannot be materially reduced below the proportion of one equivalent to one equivalent of the base, under the new conditions the mineral acid may be wholly or partially replaced by an organic acid such as formic acid, without the quality of the black being materially affected. As suitable oxygen carriers the chlorides of copper have been found to give the best results, it being preferrable to use the copper in the form of a cuprous salt. This is effected by adding to the dye mixture cupric chloride, together with a sulphite or bisulphite in sufficient quantity to reduce the cupric salt to the cuprous state, and a sufficient quantity of a soluble chloride to keep the cuprous chloride in solution. Among the compounds suitable for the production of this black in conjunction with aniline are, para-phenylene-diamine, dimethyl-para-phenylene-diamine, para-amido-diphenylamine, para-amido-phenol, etc."

This method may be used alone as the other Aniline Blacks, or the dyed skins may be after-dyed in a bath containing a logwood dye, or it may be used in conjunction with mineral dyes, or with the Oxidation Colors (see next chapter). A typical formula for the black by Green's process is the following:

Para-amido-phenol	0.5 kg.
Aniline oil	10 liters
Hydrochloric acid 22° Bé.	10 liters
Acetic acid 40%	5 liters
Cold water	25 liters

This is solution A. Solution B is prepared by dissolving

Copper sulphate   2 kg.

Salammoniac   10 kg.

Cold water   50 liters

A and B are mixed, and the mixture applied to the hair of the furs several times, drying each time at 35°–40° C. After three coats of dye have been applied, a pretty and fairly intense black shade is obtained, which is developed further by treating with a solution of 25 grams of sodium bichromate per liter of water. The skins are then allowed to dry in air, and then if desired, an after-dyeing is made with some other dye.

On account of its extreme fastness, Aniline Black, produced by any of the methods outlined above, has attained a justifiable popularity for the dyeing of furs, in spite of the necessity of using the more or less cumbersome brush method of applying the dye. Very recently there was issued to a German company a patent in which is described a method whereby furs can be dyed with Aniline Black by the dip process. An abstract of the patent (D. R. P. 33402) is as follows: "As is known, aniline salt, and similar salts, together with oxidizing agents like bichromates, chlorates, etc., cannot be used for dyeing furs by the dip process, because the strongly dissociated mineral acid is injurious to the leather. The dissociation of the acid can be reduced by adding neutral salts, like common salt, or Glauber's salt, so that good results can be obtained by dyeing in a bath of the dye mixture, the leather retaining its softness."

Thus far there have been no reports of the successful practical application of this patent, so its value cannot be discussed. It is extremely doubtful, however, that furs will ever be dyed in the dyebath with the present type of Aniline Black formulas, no matter what substances are added to prevent the leather from being affected.

those of the Aniline Black process, though possibly not so complicated, with the important difference, however, that, while in the production of Aniline Black acid is essential, in the present instance the oxidation can be carried on in neutral or even alkaline medium. On account of the character of the method used in applying the new fur dyes, the name Oxidation Colors has been given to them. Strictly speaking, Aniline Black is also an Oxidation dye, but it is usually considered in a class by itself. The methods used at first in the application of the Ursol dyes to furs followed closely the process as described in the patents. The furs were first killed, usually by brushing on a lime mixture, drying, and then beating out the dust. This operation was repeated, if necessary. Then a solution of the desired dye, mixed with an equal volume of 3% peroxide of hydrogen was brushed on and the fur allowed to lie exposed to the air. The dyeing could also be done by the dip process, less concentrated solutions being used. By varying the concentration of the solution, and prolonging or shortening the time of action, the shades could be varied from very light to very dark, and by combining two or more of the Oxidation Colors, many different color effects could be produced. Soon other fur dyes were developed and put on the market; for example, Ursol DB, giving blue to blue-black shades, and Ursol 2G, yielding yellowish tones suitable for mixing with the other colors. Ursol C was discarded shortly after its introduction. The dyeings obtained with the Oxidation Colors seemed to be very fast, resisting successfully the action of cold or hot water, or even hot soap solution. Moreover, a dyed hair examined under the microscope appeared to be colored through the epidermis to the medulla, and no individual particles of dye could be discerned.

The new fur dyes had many evident advantages over the coloring matters in general use at the time. The simplicity of the dyeing operations, the short duration of the process, the great tinctorial power of the new products, were facts which strongly recommended themselves to the progressive fur dyer. The cost of the dyes was higher than that of the vegetable dyes, but this consideration was largely overbalanced by the saving in time and labor in using them. And yet, the Ursol dyes found only a comparatively small market. The majority of fur dyers, always conservative and reluctant to turn from the traditional ways of the industry were skeptical of, and even hostile towards the new dyes and the new methods of dyeing. In a sense, this opposition was justifiable. It was not an easy task to relinquish all at once methods which had been successfully applied for generations back, and with which they were thoroughly experienced, in favor of processes which were radically different, and with which they had no experience at all. But some enterprising spirits among the fur dyers undertook to try out the new products and it was not long before the skeptics had good cause for condemning the work and achievements of the chemists as far as fur dyeing was concerned. The new type of dyes did possess some of the advantages claimed for them, but they also possessed many highly objectionable features, which had never been manifest with the vegetable dyes. First of all, the dyeings were not so fast as had at first appeared, for the color came off the hair when the furs were rubbed, brushed or beaten. Then it was observed that after a short time some of the dyeings changed color, and at the same time the hair lost its gloss and became brittle. The condition of the leather after dyeing was anything but satisfactory. Most serious of all, however, was the appearance among the workers in the dyeing establishments, and also among the furriers who worked with the dyed skins, of certain pathological conditions which had hitherto been unknown. Various skin diseases, eczemas, inflammation of the eyes, asthmatic affections and intestinal irritations were some of the afflictions which were directly attributable to the use of fur dyes of the Ursol type. Medical science was at a loss to know how to treat these ailments, because their nature was not understood.

Here indeed, were obstacles threatening to destroy all the hopes which the discovery of the new class of dyes had aroused, and to check at the outset the possibility of rational progress in the fur dyeing industry. But the men of science were not content to let the matter drop thus. Difficult problems had been solved before, and surely there must be some way of overcoming the objections and deleterious features of a system of fur dyeing which had so much potential merit. Where hindrances sprang up in the path of progress, it was the duty of the chemist to remove them, and when difficulties arose, it was up to him to resolve them, as far as was humanly possible. So the chemists who had been responsible for the introduction of the Oxidation Colors set themselves to the task of eliminating the undesirable or injurious qualities. It was many years before the results of painstaking effort and persistent study cleared up the causes of all the objectionable aspects of the fur dyes, and suggested means of overcoming them satisfactorily. The work had been directed to the improvement of the dyes and of the methods of dyeing with them. Purer intermediates were produced, and more easily soluble ones, so that there would be no possibility of ultra-microscopic particles of the dye being deposited on the surface of the hair from the dye solution, instead of being taken up within the hair fibre. It was this superficial deposition of minute crystals of the dye or of the only partially oxidized intermediate, on the hair, crystals so fine as to be invisible in the ordinary high-power microscope, which caused the color to come off when the furs were brushed or beaten, giving rise to a dust which was frequently very injurious to the health. Then, mordants were adopted to help fix the dyes, compounds of copper, iron, and chromium being used as formerly with the vegetable dyes, and the range of shades was also increased thereby. Certain of the Oxidation Colors had a tendency to sublime off the hair, so the dyed hair was chemically after-treated in such cases to prevent this. The causes of the pathological aspects of dyeing with the Oxidation fur dyes were not so readily disposed of. But the adoption of devices to prevent the formation and circulation of dust during the handling of the dye, the employment of adequate protection against contact with the dye or its solutions, the use of the most dilute solutions possible in dyeing, the thorough washing of the dyed skins to remove any excess of the coloring matter, the prevention of dust formation in the drying of the skins, and the rigid observance of, and adherence to hygienic laws, were all factors in the elimination of the health-impairing phases of dyeing with the Oxidation Colors.

It was only after all these improvements had been accomplished that the fur dye intermediates began to acquire a degree of popularity among fur dyers, and strange as it may seem, there was a more ready market for these dyes in America, than in Germany where they were manufactured. Other manufacturers of coal-tar intermediates also began to produce fur dyes, and so, in addition to the Ursols, there were the Nako brand, the Furrol brand, the Furrein brand, and one or two others. New dyes were invented, until the whole range of colors suitable for fur dyeing had been produced. The black dye, however, presented some difficulty. A black dye which would rival logwood blacks could not be attained. Ursol DB in conjunction with Ursol D was being used to produce bluish-blacks, but the dyeings were not fast, turning reddish after a time. In 1909, a patent was taken out for a dye mixture, which was made up like the DB brand, but instead of using toluylene diamine with para-phenylene-diamine, the new dye was made up of a methoxy, or ethoxy-diamine with para-phenylene-diamine, and it yielded brilliant bluish-blacks, which were fast, and which very nearly approached the logwood black in luster, intensity, and bloom. For some purposes, however, the production of a black color is still dependent on the use of the logwood dye.

When the Great War cut off to a large degree the importation of skins dyed in Europe, the American fur dyeing industry developed tremendously, and in a comparatively short time was able satisfactorily to accomplish in the way of dyeing furs, what had taken foreign dyers a much longer period to attain. It had been previously considered that furs could be dyed properly only by European fur dyers, but the achievements in this direction by Americans fully dispelled this belief. But the success of the fur dyers in America might not have been so marked or rapid, had it not been for the work of the American chemists. The war had also shut off the supply of German dyes, upon which the dyeing industries of America had formerly been dependent, so enterprising chemists in this country undertook to fill the need, and in a surprisingly short time, American fur dyes, in every respect the equal of the foreign product were offered to the American fur dyers, and at the present time, the requirements of the fur dyeing industry in this country are being adequately met by domestic producers. Among the brands on the market are the Rodol, Furamine, Furol, and several others. The Oxidation Colors are now being offered in a high state of purity, and easily soluble, free from any poisonous constituents, and there is absolutely no reason for the appearance of any pathological conditions among workers on dyed furs, or users of such furs, provided the necessary precautions have been taken in the dyeing process. The occurrence of any affection which can be traced to dyed fur, cannot possibly be due to the dye itself, but to gross carelessness and negligence in dyeing, and in any such event, the dyer responsible should be brought to account.

In order to get a better understanding of the nature and action of the Oxidation Colors, a typical one will be studied in some detail. The most important one in this class is para-phenylene-diamine, usually designated by the letter D in all commercial brands of this fur dye, while its chemical formula is represented as $C_6H_4(NH_2)_2$. When pure it occurs in colorless, crystalline lumps, which rapidly turn brown when exposed to the air; the technical product of commerce is of a dark-brown color. It dissolves readily in hot water when pure, and also in acids. At one time the hydrochloride was used instead of the free base, on account of its greater solubility, but now a base is made which is sufficiently pure to be very soluble in water. There are several methods of preparing para-phenylene-diamine: first, by the reduction of amido-azobenzol, the product obtained in this way always containing a slight amount of aniline, which reduces the solubility, and also gives rise to poisonous oxidation products during the dyeing process; second, by the reduction of paranitraniline, the quality and solubility of the product in this case depending on the purity of the starting material; and third, by the treatment of para-dichloro-benzol with ammonia under pressure, the best product being obtained by this method. The crude para-phenylene-diamine, made by any of the above processes, is generally distilled in vacuo, the refined base being obtained as lumps with a crystalline fracture.

The first step in the oxidation of the para-phenylene-diamine is the formation of quinone di-imine, $NH:C_6H_4:NH$. This is a very unstable compound in the free state, and even in aqueous solution it decomposes within a comparatively short time, or combines with itself to form a more stable substance. Quinone di-imine has a very sharp, penetrating odor, and produces violent local irritations wherever it comes in contact with the mucous membrane. If a small quantity of para-phenylene-diamine is absorbed into the human body, by breathing the dust, or otherwise, the formation of quinone di-imine takes place internally with consequent irritation of the mucous lining throughout the body. The various pathological conditions mentioned before may be ascribed to irritation caused by quinone di-imine. In any dyeing process where there is a possibility of the formation of quinone di-imine, as is the case with most dyes containing para-phenylene-diamine, special precautions must be taken by the workers in handling the dye or coming in contact with its solutions, and no one who is particularly sensitive to irritation should be permitted to work in a place where such dyes are used.

The next step in the oxidation of the para-phenylene-diamine is the formation of what is called Bandrowski's base. Three parts of the quinone di-imine combine with themselves, forming a substance of a brown-black color, which was formerly regarded as the final oxidation product. The formula of Bandrowski's base is represented by the following chemical hieroglyphics:

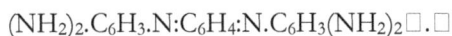

$(NH_2)_2.C_6H_3.N:C_6H_4:N.C_6H_3(NH_2)_2 \square . \square$

Further investigation has shown that the oxidation proceeds beyond this stage with the formation of a compound of what is known as the azine type, which is depicted by the chemist as

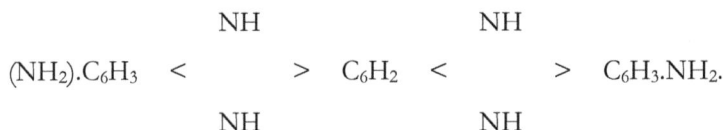

$$
(NH_2).C_6H_3 \quad <{\displaystyle {NH \atop NH}}> \quad C_6H_2 \quad <{\displaystyle {NH \atop NH}}> \quad C_6H_3.NH_2.
$$

It is by no means certain that this substance is the true coloring matter obtained by the oxidation of para-phenylene-diamine, for the reactions may continue still farther, producing even more complicated oxidation products. Scientific research and study has not as yet gone beyond this stage.

The reactions of the other dyes of the Oxidation type are quite similar to those of para-phenylene-diamine, some being simpler, and others being even more complex. The presence of certain chemical groups in the intermediate, or the relative position of such groups are factors responsible for the variations in shade.

With the various mordants, the Oxidation Colors give different shades, and a great range of colors can be produced either by combining mordants, or combining dyes, or both. The following tables illustrate the shades formed with the customary mordants.

	CHROME	COPPER	IRON	DIRECT
Ursol D	brown black	coal black	coal black	dark brown to

				brown black
Ursol P	dull red brown	dull dark brown	grey brown	light brown
Ursol 2G	yellow brown	dull yellow brown	yellow brown	dull yellow
Ursol A	...	...	blue black	blue to blue-black
Ursol 4G	light brown	medium brown	yellow	pure yellow
Ursol 4R	orange brown	light yellow brown	red brown	orange red
Ursol Grey B	greenish grey	greenish grey	mouse grey	...
Ursol Grey R	brownish grey	brownish grey	reddish grey	...

Fur dyes of American make being equal in every way to the German product, show the same color reactions with the various mordants. The following table shows the shades produced with the same mordants as above:

	CHROME	COPPER	IRON	DIRECT
Rodol D	brown black	coal black	coal black	brownish black
Rodol P	red brown	dark brown	grey brown	light brown
Rodol 2G	yellow brown	yellow brown	yellow brown	dull yellow
Rodol 4G	light brown	light brown	reddish brown	pure yellow
Rodol A	...	blue black	...	blue black
Rodol Grey B	greenish grey	greenish grey	mouse grey	...
Rodol Grey R	greenish grey	brownish grey	mouse grey	...

All these shades are produced by dyeing in a bath containing a *neutral* solution of the dye. Sometimes the dye comes in the form of a salt of a mineral acid, like hydrochloric or sulphuric acid, in which case a sufficient amount of an alkali, usually ammonia, is added to liberate the free base. According to the Cassella Co., German manufacturers of the Furrol brand of fur dyes, the dyeing can also be carried on in slightly alkaline or in slightly acid solution, a different series of shades being obtained in each instance. Ammonia is used to render the bath alkaline, and formic acid to make it acid. The most customary practise, however, is to use neutral solutions of the dyes.

For preparing the mordant solutions much smaller quantities of the metallic compounds are used than in the case of the vegetable dyes. With chrome mordants cream of tartar is always employed as an assistant, and occasionally also with copper and with iron mordants. With copper, and also with iron mordants no addition is made at all, or sometimes a small quantity of acetic acid is added. The temperature of the mordant solution is kept about 30° C., and the duration of the mordanting varies from 2–24 hours according to the depth of shade desired. The concentration of the solution may also be varied, it sometimes being just as well to use a strong mordant solution and less duration of mordanting. Chrome may be combined with copper, and iron may be combined with copper, but chrome and iron do not go together as mordants. Some typical average mordanting formulas are as follows:

Chrome mordant.

Bichromate of soda 2.5 gms.

Cream of tartar   1.5 gms.

Water   1 liter

Copper mordant.

Copper sulphate 2 gms.

(Acetic acid 50% 2 gms.)

Water   1 liter

Iron mordant.

Ferrous sulphate 2 gms.

(Acetic acid 50% 2 gms.)

Water   1 liter

or,

Iron pyrolignite 30% 10 gms.

Water   1 liter

Chrome-copper mordant.

Bichromate of soda   2 gms.

Copper sulphate   0.25 gms.

Cream of tartar   1.0 gms.

Water   1 liter

Copper-iron mordant.

Copper sulphate 2 gms.

Ferrous sulphate 2 gms.

(Acetic acid 50% 2 gms.)

Water   1 liter

The killed skins are immersed in the mordanting solution, and allowed to remain the required length of time. They are then thoroughly rinsed to remove any excess of the mordant, and are hydro-extracted. Under no circumstances should mordanted skins be permitted to dry, for they would be unfit for use again.

The dyebath is next prepared by dissolving the necessary quantity of the dye, varying from 0.1 gm. to 10 gms. per liter. Then if the solution must be neutralized, the ammonia is added and the temperature of the bath is brought to 30–35° C. by the addition of cold water. This temperature is maintained throughout the dyeing operation. To the solution is added the oxidizing agent. Ordinary commercial peroxide of hydrogen containing 3% by weight is the usual oxidizer, although perborates have been suggested. 15–20 parts of peroxide of hydrogen for every part of dye are added, and the dye solution brought to the proper dilution. As soon as the dyebath is ready, the skins are entered, and worked for a short time to effect even penetration. They are then left in the dyebath for 2–12 hours or longer according to the depth of shade. After being satisfactorily dyed, the furs are rinsed thoroughly, hydro-extracted and dried and finished. Where the dye is to be applied by the brush to the tips of the hair, stronger dye solutions are used, the brushed skins being placed hair together and let lie for about 6 hours in order to permit the color to develop, after which the furs are dried and drum-cleaned.

Some shades, particularly black, have a tendency to rub off slightly. In order to overcome this, the dyed furs, after rinsing, are treated with a cold solution of ½ part of copper sulphate per 1000 parts of water, for 3–4 hours, then without rinsing, hydro-extracted and dried. Furs which have been tipped are brushed with a 1–2% solution of copper sulphate and dried. Care must be taken in this after-treatment, for the use of too strong a solution of copper sulphate, or too prolonged action of such a solution will materially alter the shade of the dyed fur.

A few typical formulas will serve to illustrate the general methods of employing the Oxidation Colors:

Brown Sable Imitation on Unsheared Rabbit

The skins are killed with soda, soured, and washed, then mordanted with

Bichromate of soda   2 grams

Copper sulphate   .25 grams

Cream of tartar   1 gram

Water   1 liter

for 24 hours. Then washed, and dyed for 24 hours with

Fur Brown 2G   3 grams

Hydrogen peroxide 45 grams

Water   1 liter

Wash and dry the skins, then brush the tips with

Fur Brown D   20 grams

Hydrogen peroxide 400 grams

Water   1 liter

Black on Sheared Muskrat

The skins are killed with soda, soured, and washed, then chrome mordanted for 6 hours. Then they are dyed for 6 hours with

Rodol P	1.5 grams
Pyrogallic acid	.7 grams
Ammonia	2.0 grams
Hydrogen peroxide	45 grams
Water	1 liter

The dyed skins are washed and dried, then tipped with

Rodol D	20 grams
Rodol DB	2 grams
Hydrogen peroxide	450 grams
Water	1 liter

Brown on Thibet Sheep Skin

The killed skins are mordanted for 6 hours with a chrome mordant, then dyed for 6 hours with

Ursol P	1 gram
Pyrogallic acid	1 gram
Ammonia	2 grams
Hydrogen peroxide	40 grams
Water	1 liter

It is also possible to combine dyeings with the Oxidation Colors with Vegetable dyeings, or with Aniline Black. For example, if it be desired to produce an imitation skunk on a raccoon, and an exceptionally fast and intense and lustrous black on the tips of the hair, the skins are dyed in the bath with the Oxidation dyes, and the tips of the hair are brushed with a mixture such as described under Vegetable Colors for the production of French seal, as follows:

Imitation Skunk on Raccoon

The skins are killed with caustic soda, soured and washed, then mordanted with an iron-copper mordant as described, and then dyed with

Fur Grey R	3 grams
Ammonia	2 grams
Peroxide of hydrogen	45 grams
Water	1 liter

After washing and drying, the dyed skins are brushed over with a mixture such as used for dyeing French seal with Vegetable Colors.

In a similar manner, the Oxidation Colors may be used to give a base color to furs dyed by the Aniline Black process.

It is apparent from these few illustrations that a great variety of shades can be produced, and the dyeing of imitations of the better class of furs on cheaper skins is a comparatively simple matter, after an understanding of the nature of the dyes has been obtained, and a certain amount of skill acquired in working with these dyes.

CHAPTER XVI
FUR DYEING
COAL TAR DYES

In addition to the Aniline Blacks and the Oxidation Colors already discussed there are certain of the synthetic coal tar dyes such as are generally used in the dyeing of textiles, which can also be applied on furs. There are several classes of these dyes, varying somewhat in their nature, and consequently in their manner of application; in the main they produce bright shades, such as are but seldom used on furs, yet which may occasionally serve for the production of novel effects. Basic, acid and chrome colors are the types which can be employed.

Basic colors possess great fullness and tinctorial strength, but have a tendency to rub off, and the tips of the hair take a darker shade with these dyes than the rest of the hair. The addition of acetic acid and Glauber's salt to the dyebath will result in a more uniform dyeing. On account of the comparatively poor fastness to rubbing and washing, basic dyes are used only for dyeing furs which are intended for cheap carpet rugs, such as sheep and goat. They may also find use in the production of light fancy shades on other white furs. The procedure is usually as follows: The furs are killed in the customary manner with soap and soda or ammonia, or if this is insufficient, with milk of lime. A soap-bath is then prepared containing 2.5–6 grams of olive-oil soap per liter of water. The temperature of the bath is brought to 40° C. To this is added the solution of the dyestuffs, prepared by mixing the required color or colors with a little acetic acid to a paste, and then pouring boiling water on the mixture until dissolved. Undissolved particles or foreign matter are removed by passing this solution through a cotton cloth or sieve, and the clear solution then mixed with the soap-bath. The well-washed skins are then entered into the dyebath and immersed for about half an hour, or until the desired depth of shade is obtained. They are then removed, pressed or hydro-extracted and dried. For the production of light shades, the following dyes may be used:

For cream, light sulphur-yellow, maize, salmon, etc.

- Combinations of
- Thioflavine
- Rhodamine B
- Irisamine G

For greenish-yellows

- Combinations of
- Thioflavine
- Victoria Blue B

For light pink

- Rhodamine B
- Irisamine
- Rose Bengal Extra N

For purple

- Methyl Violet 3B–6B
- Crystal Violet

For sky-blue

- Victoria Blue B

For white

- Victoria Blue B (Milk-white)
- Methyl Violet 3B–6B
- Crystal Violet (Ivory-white)

To produce very delicate shades, the moist dyed skins are subjected to a sulphur bleach overnight, to lighten the color, then rinsed, and dried. Full, brilliant shades may be obtained by dyeing in a bath of 40° C., acidulated with 2–3 grams of acetic acid per liter of solution, the following dyestuffs being suitable:

For yellow to orange

- Thioflavine
- Paraphosphine
- Rhodamine
- Safranine
- New Magenta O

For pink

- Rhodamine B
- Rose Bengal Extra N

For light red

- Safranines

For bordeaux and red

- Magenta
- New Magenta
- Russian Red
- Cerise

For violet

- Methyl Violet 6B–4R
- Crystal Violet 5B

For blue

- Victoria Blue B
- Methylene Blue BB
- New Methylene Blue N

For green

- Malachite Green Crystals
- Brilliant Green Crystals, or combinations of
- Thioflavine
- Diamond Phosphine
- Victoria Blue B

For brown

- Chrysoidines
- Bismarck Browns

In dyeing skins with harder hair than that of sheep or goat, mere killing is insufficient to render the hair capable of taking up the dye. The skins are therefore immersed before dyeing, in a cold, weak solution of chloride of lime, the affinity of the hair for the dye being thereby greatly increased.

Acid dyes are employed when a greater fastness is required than can be obtained with the basic colors. Sulphuric acid in a quantity equal to half the weight of the dyestuffs used, together with four times that quantity of Glauber's salt is added to the dyebath. Formic acid may be used in place of the sulphuric acid, very good results being obtained. The skins are immersed in the dyebath, and worked until thoroughly soaked with the dye liquor, and then allowed to remain until the proper depth of shade is attained, or overnight. The temperature of the solution is about 40° C., and only very light shades can be produced in this manner. In 1900 and again in 1914, the Cassella Co., a large German manufacturer of dyestuffs, obtained patents for processes enabling the dyeing of furs in hot solution with the acid dyes. The method required that the skins be chrome-tanned in order to render them resistant to the action of hot solutions, the addition of a small amount of formaldehyde to the chrome solution increasing this effect. The skins are then treated with a solution of chloride of lime in order to increase the affinity of the hair for the dyestuffs. The method as it is now practised is as follows: The skins which have been cleaned and washed are chrome tanned by the method as described in the chapter on Tanning Methods, 60 grams of formaldehyde being added to every 10 liters of the chrome solution. After proper tanning the skins are rinsed, and while still moist they are subjected to a treatment with chloride of lime. They are first immersed for 15 minutes in a cold bath containing 120 grams of hydrochloric acid 32–36° Twaddell per 10 liters of water, then without rinsing, they are entered into a bath made up by adding gradually in four portions the clear solution of 2–4 grams of the chloride of lime per 10 liters of water. After working for an hour, the skins are removed and entered again into the acid solution, in which they are worked for another 15 minutes. In order to neutralize and remove the last traces of the chloride of lime from the furs, they are rinsed in a luke-warm bath containing 1–2 grams of sodium thiosulphate, or hyposulphite of soda, in 10 liters of water. The skins are then rinsed again, and hydro-extracted, or pressed, and are ready for dyeing. The dyebath is prepared with the required quantity of dye, to which is added 10–20% Glauber's salt and 2–5% acetic acid (both calculated on the weight of the skins). The skins are entered at 20° C., then after three-quarters of an hour to 40° C., and then after another hour slowly to 50–55° C. For blacks, the temperature is raised as high as 65° C. After dyeing the skins are treated with a solution containing per 10 liters

90–120 grams of olive-oil soap

12–25 grams olive oil

12 grams ammonia

for 15 minutes, then hydro-extracted and dried, without further rinsing.

For this method of dyeing, the following dyes may be used:

For yellow and orange

- Fast Yellow S
- Acid Yellows
- Naphthol Yellow S
- Tropaeoline
- Orange GG, R, II, IV

For reds

- Acid Reds
- Lanafuchsine
- Azo Orseille

For violet

- Azo Wool Violet
- Acid Violets

For blue

- Cyanole FF
- Azo Wool Blue
- Naphthol Blue R
- Formyl Blue B

For green

- Naphthol Green B
- Fast Acid Green
- Cyanole Green

For brown, combinations of

- Fast Yellow S
- Acid Yellows
- Tropaeoline DD
- Orange GG

- Lanafuchsine

- Indigo Blue N

- Cyanole B

- Fast Acid Green BN

For black

- Naphthylamine Blacks

- Naphthol Blacks

- Naphthol Blue-black

For grey

- Silver Grey N

- Dyed with the addition of ½–1% of alum

The chrome colors are dyed on furs when very fast shades are desired, all the fancy colors being produced in this manner, but for black, only the acid dyes are suitable. The preparation of the skin is exactly the same as for the acid colors, except that the treatment with chloride of lime may be omitted, although for very full shades it is desirable. The dyeing is carried out as follows: The dyebath is prepared with the requisite amount of the desired dyestuff, which is previously dissolved, and to this is added a solution of sodium bichromate, the amount of this substance being half the weight of the dye. The solution is heated and the skins entered and dyed for 1–2 hours at 70–80° C. Then the dyebath is exhausted by the addition of ⅓% acetic acid, the skins being worked for another half hour, then rinsed, hydro-extracted and dried. Any of the one-bath chrome, or after-chrome colors may be used for this method.

Recently methods have been patented for the dyeing of furs by means of the vat colors. Vat dyes are among the fastest coloring matters ever produced, and their application on furs would be a great advantage, if suitable shades could be obtained. The general process for dyeing with vat colors, consists in reducing the dye, which is usually very insoluble, into a soluble "leuco" compound, by means of hydrosulphites in the presence of alkalies. The leuco compound is not a dye itself, but when the fibre absorbs it, and is then exposed to the air, the leuco compound is reoxidized to its original insoluble form, which remains fast and permanent. The use of strong alkalies in vat dyeing has hitherto been a great obstacle in the use of these dyestuffs, but in 1917, the Farbwerke Hoechst, a large German dye works, patented a process as follows: "A process for dyeing furs with vat colors. The dyeing is done in solutions of the vat dyes (after the addition of gelatine or some other protective colloid), which are rendered neutral or only slightly alkaline with ammonia, by neutralizing the caustic soda of the solution of the leuco compound of the vat dyes by the addition of ammonium salts, or suitable acids. The dyeings thus obtained are uniform and fast, the leather is dyed to only a slight degree, and shows no deleterious effects of the dyebath on the tannage." As a practical application of this process, another patent was taken out by the same company, also in 1917, as follows: "A process for producing fast blacks on furs, consisting of dyeing a ground color with appropriate vat dyes in a hydrosulphite vat, and after oxidation in air, topping with an Aniline or Diphenyl black. The dyeings obtained by the combination of vat dyes which are fast to oxidizing agents, with an oxidation black, have an appearance matching that of logwood black in beauty; and with a dark-blue to blue-black under-color, and a full, deep black top color, cannot be distinguished from logwood. These dyeings also have the advantage of being faster to light than logwood or other blacks."

While these processes undoubtedly have many meritorious qualities which make them interesting, they do not seem as yet, to have attained any great practical application. However, it is a field of fur dyeing which is worth while developing, and with certain necessary improvements in these processes, the vat dyes may yet supersede partially some of the other methods of dyeing furs.

CHAPTER XVII
BLEACHING OF FURS

Bleaching is for the purpose of lightening the color of furs, and is most generally applied to white-haired skins such as white fox, ermine, and occasionally white lambs of all kinds, and white bears. Among such furs, pelts of a naturally pure white tone are relatively scarce, while in the majority of cases the color ranges from a pale creamy white to a decidedly yellowish shade. Colors which vary from the pure white detract considerably from the attractiveness and consequent value of the fur, and indeed, some pelts are so far off shade that they can only be used when dyed a darker color. Most white skins which are but slightly inferior in color can be brought to a pure white by bleaching, and they can then be used natural. Some pelts, on the other hand, are particularly resistant to the action of bleaching agents and cannot be sufficiently decolorized to render them suitable for use natural, so these are also dyed. For the production of certain delicate or fancy dyed shades on white furs, it is often necessary to bleach the skins in order to be able to obtain pure tones. Such instances are not very common, however. Occasionally dark furs, such as beaver, are bleached on the tips of the hair, a golden shade being obtained thereby, which at one time was quite popular, but recently such effects have not been in vogue.

In the bleaching of furs, two steps may be distinguished, first degreasing, and second, bleaching proper. In the preliminary operations of fur dressing, the furs are treated with soap or weak alkalies to cleanse them and to remove excess oil from the hair. During the various processes and manipulations, the hair, especially on white skins, may become soiled or somewhat greasy again, so it is advisable to repeat the cleaning process. This should in every case be as light as possible, using a weak solution of soap for the softer and cleaner pelts, or dilute solutions of ammonium carbonate or soda ash for the more greasy-haired skins. The skins are then thoroughly rinsed to remove all traces of the degreasing material. This step is very essential in order to obtain uniform bleaching.

Broadly speaking, there are two general methods which can be used in bleaching furs, one involving the use of what are known as reducing agents, and the other employing oxidizing substances.

Among reducing agents which can be used for bleaching furs are sulphurous acid, and its salts such as sodium bisulphite and sodium sulphite; hydrosulphites, and derivatives.

1. **Sulphurous acid.**—When sulphur is burned, sulphur dioxide gas is formed. In the presence of moisture, or when dissolved in water, this gas forms sulphurous acid, which is one of the most commonly used bleaching chemicals for all sorts of materials, and is very effective for decolorizing furs. The procedure usually followed is to hang up the moistened skins on wooden rods in a more or less cubical chamber made of stone or brick, and lined with wood or lead. No other metals may be used, because they are quickly corroded by the sulphurous acid. The requisite quantity of sulphur is placed in a pot in the bleaching chamber, and then ignited, after which the doors are shut tight. The fumes of the burning sulphur in contact with the moist hair readily exert their bleaching action on the furs, and the operation is allowed to proceed for six or eight hours, or overnight. Then by means of fans or other devices, the air filled with sulphur dioxide gas is withdrawn from the chamber, and replaced by fresh air. The door is opened, the skins removed, exposed to the air for a time, then rinsed, and finally dried and finished. Sometimes one operation is not enough to sufficiently bleach the hair, so the process is repeated. Sulphur dioxide gas can now be obtained compressed in cylinders, which are more convenient to handle than burning sulphur. The flow of gas which is introduced into the bleaching chamber by means of a nozzle attached to the cylinder, can be regulated, and the bleaching thus retarded or accelerated.

2. **Sodium bisulphite and sodium sulphite.**—These salts of sulphurous acid are effective in their bleaching action only when in solution in the presence of acids. The acids liberate sulphurous acid from the salts, so this method is virtually the same as 1. Instead of using the salts of sulphurous acid, sulphur dioxide may be dissolved in water, and the solution used for bleaching by immersing the furs in it. This procedure, while consuming somewhat less time than the chamber process, is more likely to affect the leather, which would have to be retanned. The principle is the same as that involved in method 1.

3. **Hydrosulphites and derivatives.**—The bleaching agent can be prepared by adding zinc dust to commercial bisulphite of soda dissolved in about four times its weight of water until no more reaction is evident. Milk of lime is then added to precipitate the zinc, and the clear supernatant liquid of 1.5°–5° Tw. is used for bleaching. The skins are immersed for 12–24 hours, taken out, washed and finished. Instead of preparing the hydrosulphite, the commercial products may be used with greater convenience, a solution containing 1–4% of the hydrosulphite powder being used, and the skins treated in this until satisfactorily bleached.

The bleaching action of sulphurous acid and hydrosulphite is supposed to be due to the reduction of the coloring matter of the hair to a colorless compound; or possibly to the formation of a colorless compound of the bleaching material with the pigment. The former seems the more probable explanation, because the change is not a permanent one, the original natural color returning after a long exposure of the bleached fur to air and light. However, the results are sufficiently enduring to satisfy the requirements of the trade in the class of furs on which these methods of bleaching are used.

Bleaching chemicals with an oxidizing action generally used for decolorizing furs are hydrogen peroxide and peroxides; occasionally hypochlorites and permanganates are also used.

1. **Hydrogen peroxide.**—Hydrogen peroxide is usually employed for bleaching in the form of its 3% solution, to which is added about 20 cubic centimeters of ammonia per liter. The ammonia serves partially to neutralize the acid which commercial peroxide generally contains, and also to facilitate the bleaching action. The thoroughly degreased skins are immersed in the solution until the hair is completely wetted by it, are then removed, and evenly pressed or hydro-extracted, after which the pelts are hung up to dry in the air. As the hair becomes drier, the concentration of the peroxide becomes greater, and consequently the bleaching action is stronger. Where there is a likelihood of the leather being affected by the bleaching solution, the ammoniacal peroxide may be applied to the hair with a fine sponge or brush until sufficiently wetted, and then hanging the skins up to dry. Repetition of the process is sometimes necessary to obtain pure white, but the results are always excellent.

2. **Peroxides.**—The most important of these is sodium peroxide, which comes on the market as a yellowish-white powder, which must be kept dry, and away from any inflammable material, as fires have been caused by the contact of the peroxide with such substances. When dissolved in water, it is equivalent to a strongly alkaline solution of peroxide of hydrogen.

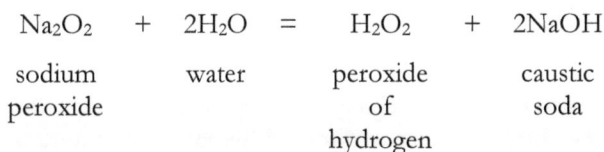

$$Na_2O_2 \quad + \quad 2H_2O \quad = \quad H_2O_2 \quad + \quad 2NaOH$$

| sodium peroxide | water | peroxide of hydrogen | caustic soda |

When dissolved in acid, the alkali is neutralized, and a neutral solution of peroxide of hydrogen and a salt is obtained, and this method is used to obtain peroxide of

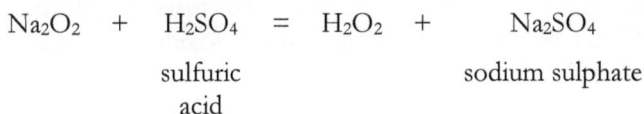

$$Na_2O_2 \quad + \quad H_2SO_4 \quad = \quad H_2O_2 \quad + \quad Na_2SO_4$$

| sulfuric acid | | sodium sulphate |

hydrogen cheaply. 3 parts of sodium peroxide are slowly dissolved in a cold 1% solution of 4 parts of sulphuric acid, stirring during the addition, and making the resulting solution neutral to litmus paper, acid or more sodium peroxide being added as needed. There is then added 3–6 parts of a solution of silicate of soda of 90° Tw. The skins are immersed until properly bleached, taken out, passed through a weak acid solution, then washed and finished. This method generally requires the leather to be retanned after bleaching. Another process, which involves the use of peroxides, but which is not commonly practised, consists in rubbing the hair with a pasty mixture of equal parts of water, barium dioxide, and silicate of soda, hanging up the skins to dry, and then beating and brushing the hair.

3. **Permanganates.**—The only member of this group that finds practical application for bleaching purposes is potassium permanganate. The skins are immersed in a 0.1% solution of the crystals of potassium permanganate, until the hair acquires a deep brown color. They are then removed, rinsed, and entered into a second bath containing sulphurous acid in solution, prepared by acidifying a solution of sodium bisulphite. The skins are then worked in this until fully bleached. It is the permanganate which does the bleaching, the sulphurous acid being for the purpose of dissolving the brown compound of manganese formed on the hair.

4. **Hypochlorites.**—Chloride of lime and sodium hypochlorite, which is prepared from the former, are the chief chemicals of this type used for bleaching. The skins are entered into a weak solution of the hypochlorite, and left until the hair is decolorized; then after removing, they are passed through a dilute acid, and subsequently through a weak solution of sodium thiosulphate in order to remove all traces of the hypochlorite. This method causes the hair to acquire a harsh feel, and the yellow color is never entirely eliminated. The hair, however, possesses a great affinity for certain types of dyestuffs, and it is only when these particular classes of dyes are to be applied to the furs, that the hypochlorite bleach is used. (See dyeing with Acid colors).

The various oxidation methods of bleaching are supposed to change the coloring matter of the hair into an entirely different and colorless compound which cannot return to its original form. The bleach is therefore permanent.

In common practise, the sulphurous acid, and the peroxide of hydrogen methods are the two chiefly employed in bleaching processes. Sulphurous acid is used to bleach the cheaper kinds of furs, while peroxide of hydrogen is applied to the finer furs.

Whichever process is used, it is customary to give the bleached skins a subsequent "blueing," by passing them through a very weak solution of a blue or violet dye, such as indigo-carmine, crystal violet, alkali blue or ultramarine. The furs are then dried and finished off as usual. In drum cleaning white furs, gypsum or white sand, or sometimes even talc are used with the sawdust, or occasionally alone without the sawdust.

BIBLIOGRAPHY

- Allen "Commercial Organic Analysis"

- Armour, B. R. "Fur Dressing and Dyeing" 1919

- —— Color Trade Journal, Vol. 1, p. 51–53

- —— Jour. Amer. Leather Chemists' Assn., Vol. 13, p. 63–69.

- Belden, A. L. "Fur Trade in America" 1917

- Beltzer, F. J. G. "Industrie des Poils et Fourrures, etc." 1912

- —— Revue Generale des Matieres Colorantes, Vol. 12, 1908

- Bennett, H. G. "Manufacture of Leather" 1910

- Bertram, P. Deutsche Färber-Zeitung 1895–96 Heft 17, p. 266

- Bird, F. J. "American Practical Dyers' Companion" p. 241–245

- Boerner, H. Kunststoffe, 1912 p. 223

- Brevoort, H. L. "Fur Fibres as shown in the Microscope" 1886

- Bucher, B. "Geschichte der technischen Künste" 1875–1893

- Cubaeus, P. "Das Ganze der Kürschnerei" 1912

- Davis, C. T. "Manufacture of Leather"

- Erdmann, E. Deutsche Färber-Zeitung 1894–95 Heft 21, p. 337

- —— Zeitschrift für angewandte Chemie, 1895, Heft 14

- —— Zeitschrift für angewandte Chemie, Heft 35, 1905

- —— Berichte, 1904, 37, p. 2776, 2906

- Farrell, F. J. "Dyeing and Cleaning" 1912

- Fougerat, L. "La Pelleterie dans l'antiquité, la préhistoire, etc."

- Fleming, L. "Practical Tanning" 1916

- Gardner, W. M. "Wool Dyeing" 1896

- Grandmougin, E. Zeitschrift für Farben-Industrie, 1906, 5, p. 141

- Gruene, E. Deutsche Färber-Zeitung, 1895–96 Heft 13, p. 197

- Halle "Werkstätte der heutigen Künste," 1762, Vol. 2, p. 317

- Hartwig, O. L. "Sprengler's Künste und Handwerke," 1782

- Hausman, L. A. Scientific Monthly, Jan. 1920; March, 1921

- —— Natural History, Vol. 20, 4, 1920

- —— American Journal of Anatomy, Sept. 1920

- —— American Naturalist, Nov.–Dec. 1920

- Hayes, A. H. National Cleaner and Dyer, Nov. 1920, p. 55–57

- Jacobson, "Schauplatz der Zeugmanufacturen" p. 493

- Jones, J. W. "Fur Farming in Canada" 1913

- Knecht, Rawson & Loewenthal "Manual of Dyeing" 1916

- Kobert, R. "Beitrage zur Geschichte des Gerbens und der Adstringentien" 1917

- Koenig, F. Zeitschrift für angewandte Chemie, 1914, Vol. 1, p. 529

- Lamb, M. C. "Dressing of Leather" 1908

- —— Jour. Soc. Dyers & Colourists 1913, 29, p. 160–165

- Lamb, J. W. Jour. Soc. Dyers & Colourists Dec. 1905, p. 323

- Larish & Schmid "Das Kuerschner Handwerk" 1–3

- Laut, A. C. "The Fur Trade of America" 1921

- Lightfoot, J. "The Chemical History & Progress of Aniline Black" 1871

- Mairet, E. M. "A Book on Vegetable Dyes" 1916

- Martin, G. "Industrial Organic Chemistry"

- Matthews, J. M. "Application of Dyestuffs" 1920

- Mayer, A. "Die Färberei in der Werkstätte des Kürschners"

- Mierzinski, S. "Die Gerb und Farbstoffextrakte"

- Noelting & Lehne "Anilin-Schwarz" 1904

- Perkins & Everest "Natural Organic Coloring Matters" 1918

- Petersen, M. "The Fur Traders & Fur-Bearing Animals" 1920

- Poland, H. "Fur-Bearing Animals in Nature and Commerce"

- Proctor, H. "Leather Industries Laboratory Book"

- —— "Tanning"

- —— "Making of Leather"

- Schlottauer, E. Deutsche Farber-Zeitung 1911, Heft 20, p. 397

- —— Deutscher Färber-Kalender 1911, p. 65

- —— Leipziger Färber-Zeitung 1909, p. 441

- Schmidt, C. H. "Handbuch der Weissgerberei"

- Setlik, B. Deutsche Färber-Zeitung 1901, p. 213

- Smith, R. W. Color Trade Journal Vol. 3, Sept. 1918, p. 304–310

- —— Textile Recorder, Vol. 36, p. 292–293, Dec. 1918

- —— Revue Generale des Matieres Colorantes, Vol. 23, p. 32–36

- Stevenson, C. H. "U. S. Fish Commission Report 1902–1903", Bulletin No. 537

- Stickelberger, E. "Geschichte der Gerberei" 1915

- Strasser "Chemische Färberei der Rauchwaren" 1879

- Ullmann "Enzyklopedie der technischen Chemie"

- Villon, A. M. "Traité pratique de la fabrication des cuirs, etc." 1900

- Werner, H. "Die Kürschnerkunst" 1914

- —— "Das Färben der Rauchwaren" 1914

- Whittaker, C. M. "Dyeing with Coal Tar Dyes" 1919

- Wiener, F. "Weissgerberei" 1877

- Witt-Lehman "Chemische Technologie der Gespinst-Fasern" 1910

- Zeidler, H. "Die moderne Lederfabrikation" 1914

FOOTNOTES:

Descriptions after W. S. Parker, Deputy Chairman, Fur Section of London Chamber of Commerce, in Encyclopedia Britannica, 11th Ed.

Descriptions and figures taken from "Mammal Fur Under the Microscope," by Dr. L. A. Hausman, in *Natural History*, Sept.–Oct., 1920.

Inasmuch as most manufacturers use the same letters to designate the various dyes, any equivalent brand of fur dye may be used in place of those here mentioned.

Milton Keynes UK
Ingram Content Group UK Ltd.
UKHW030741071024
449371UK00006B/665

9 789362 516176